Ulrich Reimers

Digital Video Broadcasting (DVB)

Springer

Berlin
Heidelberg
New York
Barcelona
Hong Kong
London
Milan
Paris
Singapore
Tokyo

Ulrich Reimers

Digital Video Broadcasting (DVB)

The International Standard
for Digital Television

With 175 figures

 Springer

Prof. Dr.-Ing. Ulrich Reimers

Dr.-Ing. Frank Fechter
Dipl.-Ing. Dirk Jaeger
Dipl.-Ing. Christian Johansen
Dipl.-Ing. Uwe Ladebusch
Dipl.-Ing. Christof Ricken
Dipl.-Ing. Alexander Roy
Dipl.-Ing. Markus Trauberg
Dipl.-Ing. Andreas Verse

Technische Universität Braunschweig
Institut für Nachrichtentechnik
(Braunschweig Technical University
Institute for Communications Technology)
Schleinitzstrasse 22
38092 Braunschweig
E-mail: u.reimers@tu-bs.de

TK6678
.D5213
2001

044841632

Library of Congress Cataloging-in-Publication Data
Digitale Fernsehtechnik, English
   Digital video broadcasting (DVB) ; the international standard for digital television /
   [edited by] Ulrich Reimers.
     p. cm.
   Includes bibliographical references and index.
   ISBN 3-540-60946-6 (alk. Paper)
   1. Digital television--Standards. 2. MPEG (Video coding standard). 3. Television broadcasting.
   I. Reimers, Ulrich. II. Title.

ISBN 3-540-60946-6 Springer-Verlag Berlin Heidelberg New York

Springer-Verlag Berlin Heidelberg New York
a member of BertelsmannSpringer Science+Business Media GmbH
Printed in Germany
© Springer-Verlag Berlin Heidelberg 2001

Typesetting: Data Conversion by typetwo, Berlin
Cover-Design: Struve & Partner, Heidelberg

SPIN: 10525858   Printed on acid-free paper   68/3020 – 5 4 3 2 1 0 –

Preface

Digital Television ("Digital TeleVision Broadcasting" [DTVB] or "Digital Video Broadcasting" [DVB]) has become one of the most exciting developments in the area of consumer electronics at the end of the twentieth century. The term digital television is not typically used either to describe the digitisation of the production studio or to indicate the advent of digital signal processing in the integrated circuits used in television receivers. Rather, digital television refers to the source coding of audio, data and video signals, the channel coding and the methods for the transport of DVB signals via all kinds of transmission media. The term normally also embraces the technologies used for the return path from the consumer back to the programme provider, techniques for scrambling and conditional access, the concepts of data broadcasting, the software platforms used in the terminal devices, as well as the user interfaces providing easy access to DVB services.

The aim of this book is to describe the technologies of digital television. The description refers to a point in time at which much of the technical development work had taken place. No doubt, in the future there will inevitably be a considerable number of modifications and additions to the list of technical documents describing the technologies used for DVB. The performance data of the DVB systems, specifically of the one used for terrestrial transmission, are still in the process of being evaluated in many countries throughout the world. In some respects this book must therefore be regarded as a report on the present intermediate stage of the development of digital television and of the practical experience gained so far. I nevertheless consider it timely to present such a report, since at the date of its publication DVB will have become a market reality in many countries.

The focus is on the developments within, and the achievements of, the "Moving Pictures Experts Group (MPEG)" in the area of source coding of audio and video signals, followed by an extended description of the work and the results of the "DVB Project", the international body dealing with the design of all the other technical solutions required for the successful operation of digital television. The combination of the specifications designed by MPEG and those designed by the DVB Project has led to the overall system that we can now call digital television. The system for the terrestrial transmission of MPEG signals presented by the "Advanced Television Systems Committee (ATSC)" will not be described in detail in this book. This system will be used in the United States of America, in Canada, Mexico and some other parts of the world. It uses vestigial side-band modulation (VSB) of a single carrier per channel, and sound coding called Dolby AC-3. VSB will be explained in the chapter dealing with modulation techniques.

The book addresses readers who wish to gain an in-depth understanding of the techniques used for digital television. They should already have a good knowledge of the existing analogue television systems and should know the properties of analogue audio and video signals. They should also have some understanding of the techniques used in digital signal origination and digital signal processing.

Chapter 1 is an introduction to the DVB world. It presents an overview of the whole DVB scenario and answers the very fundamental questions concerning the goals of the development of digital television. It describes the state of technical developments and tries to evaluate several scenarios for the introduction of services using the different transmission media.

Chapter 2 recapitulates the fundamentals of the digitisation of audio and video signals. It explains the parameters chosen for the digitisation, such as the quantisation scales and the sampling frequencies, and derives the resulting data rates. On the basis of the figures presented in chapter 2 the fundamentals of source coding of audio and video signals are described in chapters 3 and 4, respectively. The primary goal of source coding is to reduce the data rate for the representation of audio and video signals in such a way that no deterioration of the perceptual quality results or, at most, only a well-defined one. In this way the resources required for the transmission and/or storage of the signals can be limited.

Before the data-reduced signals can actually be transmitted they have to be amalgamated into a system multiplex. Tools for the synchronisation of audio and video, auxiliary data needed for the description of the multiplex and of the programme content conveyed by that multiplex, teletext data, and a great deal of other information need to be added. Chapter 5 explains the multiplexing and the structure of the auxiliary data.

One of the special features of digital signals is that they can be protected against the effects of unavoidable errors occurring during transmission by adding forward error correction before the signals are sent. Chapter 6 describes methods of forward error correction (FEC) in general. It then concentrates on the two methods used in the DVB world, namely, Reed-Solomon coding and convolutional coding.

Digital modulation is dealt with in chapter 7. Here again we highlight those methods which are used in DVB (QPSK, QAM, OFDM, VSB).

The techniques used for the scrambling of digital signals are presented in chapter 8. Owing to the very nature of this topic it is impossible to describe in detail the specific methods which are used for DVB.

The three standards for the transmission of DVB signals via satellite, on cable and via terrestrial networks are described in chapters 9, 10 and 11, respectively. In addition to merely explaining the specific details of the standards, we give the performance data and further information regarding the hardware implementation in the receivers.

Finally, the methods of measuring and evaluating DVB signals as well as audio and video quality are explained in chapter 12.

This book is the result of the joint effort of several researchers working at the Institute for Communications Technology (Institut für Nachrichtentechnik – IfN) at Braunschweig Technical University (Technische Universität Braunschweig) in Germany. The contents are based upon a series of seminars held for interested European industrial professionals since the beginning of 1994. To date, over 250 participants have attended such seminars, in which we not only present the theoretical background of the DVB systems but demonstrate the possibilities of DVB and the existing DVB systems and services. In nearly all cases the authors report about areas in which they themselves have carried out research work. Amongst other things, this work has already generated four doctoral theses. Several of the researchers are – or have been – members of either ad-hoc groups of the DVB Project or of the Moving Pictures Experts Group (MPEG).

I wish to thank the authors for their competency and co-operation in meeting deadlines for the various manuscripts. Ms. Boguslawa Brandt and Ms. Simone Sengpiel prepared a large number of the drawings used in this book and Dipl.-Ing. Christian Johansen was responsible for the final co-ordinating and editing of the whole text. I extend my thanks to all three of them.

The first German edition of this book was published in 1995. It was followed by a revised second edition in 1997. The English version is based upon that second German edition. Of course, several amendments portraying developments between 1997 and 1998 have been included. The translation from German into English was done by Ms. Vivienne Bruns and Ms. Anne Kampendonk. I am most grateful to both ladies for having taken on this very difficult job.

Dr. George Waters, the former Director of Engineering of the European Broadcasting Union (EBU), undertook the final proofreading of the English version. I am especially indebted to him for his invaluable contribution to the quality of the book.

Springer publishers, and in particular Dipl.-Ing. Thomas Lehnert, were competent and active partners in producing the book. My thanks go to them for their guidance and assistance during the publication process of the English version.

Digital television, based on the specifications developed by the DVB Project, would most certainly have remained a dream without the co-operation of countless researchers and engineers in Europe, North America and Asia who were highly enthusiastic about the creation of the digital television paradigm. I have the pleasure of being chairman of the Technical Module of the DVB Project and I wish to join the other authors of this book in thanking all the DVB colleagues and friends for such a great achievement.

Braunschweig, December 1998   Prof. Dr.-Ing. Ulrich Reimers

Table of Contents

1 Digital Television – a First Summary

1.1 Definitions and Range of Application

The expression "digital" is one of the elements of language which has frequently been used in the past few years and very often misused. "Digital" only means that there are elements which can be counted on the fingers. In the everyday language of information technology the expression "digital" is often used as a synonym for "sampled, quantised and presented in binary characters". In the field of electronic media technology "digital" is a mark of quality which was first used when the compact disc (CD) was introduced on the market and offered to the general public.

The expression "digital studio" describes the introduction of digital signal forms into radio and television production. The audio and video signals are sampled by means of predetermined sampling frequencies and quantised by predetermined numbers of quantisation steps. As a rule, further source coding is not used during processing and distribution, however it may take place in individual pieces of equipment.

The signals which are received by "digital" television receivers at the moment are still analogue input signals (D2-MAC, HD-MAC, MUSE, NTSC, PAL, PALplus, SECAM), which are then internally digitised and processed as digital signals. The presentation of the image, however, remains analogue.

On the other hand, analogue "High Definition TeleVision (HDTV)", for instance "High Definition Multiplexed Analogue Component (HD-MAC)", is produced by means of completely digitised signal processing and is literally only transformed into an analogue signal immediately before transmission.

From what has been said it is evident that terminology must be carefully defined in order to limit confusion to a minimum. "Digital Television", sometimes termed "Digital TeleVision Broadcasting (DTVB)", or somewhat haplessly "Digital Video Broadcasting (DVB)", usually means the transmission of digitised audio, video and auxiliary information as data signals. However, since these data signals must in most cases be modulated onto continuous-time carrier waves in order to fit them in the transmitting channel, the actual transmission in digital television uses analogue signals.

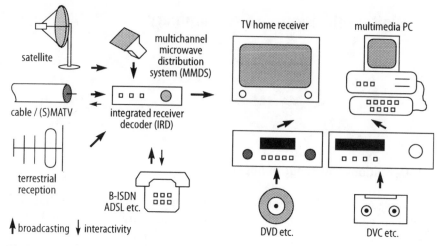

Fig. 1.1. A scenario for the utilisation of DVB

In the context of this book the expression "digital television" should in fact be understood as the system for the transmission of audio, video and auxiliary data described above. However, it is not possible to concentrate only on the channel coding and modulation, further necessary system components must be included, in particular the technology of source coding of audio and video signals and multiplexing. The latter is required for the co-ordination of various elementary streams to form one signal. The expression "multiplexing" is not really quite apt because this simplifies the actual task. In the concept of the Moving Pictures Experts Group (MPEG) the term "systems" is used to indicate that the actual multiplexing is overlaid with a large number of other tasks.

During the development of the system further elements to those named as belonging to the field of "digital television" have been more or less successfully subsumed under this expression. Figure 1.1 shows a system scenario which describes the equipment technology of digital television from the viewpoint of the user. In the main, DVB includes all equipment required for distribution, reception, signal representation and processing. However, it excludes studio technology and the actual video display.

The most suitable distribution systems for the transmission of DVB are satellite, cable, and satellite master antenna TV (SMATV). Distribution via conventional terrestrial transmission networks is also planned in several countries. One of the most innovative possibilities for digital television is the transmission of DVB signals via telephone lines. The utilisation of glass fibre, where this is available, appears to be relatively unproblematic, even for signals with a higher data rate, due to its inherent transmitting capacities. The reduction of the data rate of a television programme to just a few

Mbit/s will enable the transmission of a single television programme over a transmission distance of several kilometres via a copper telephone wire between the last switching centre and the subscriber. Finally, in many countries microwave transmission of DVB signals is exploited (Microwave Multipoint Distribution System [MMDS]) to transmit a variety of programmes, similar to those offered via a typical cable network, to private homes. The particular advantage of MMDS is the fact that, for example, thinly populated regions where a cable connection would not be financially viable can be connected.

Although satellite and terrestrial transmissions are not (yet) return-channel-compatible, a utilisation of telephone lines as well as cable networks or master antennas for return information, i.e. from viewer to network operator or content provider, is envisaged. The utilisation of a return channel for the ordering and payment of pay-TV services, or in connection with teleshopping services etc., is planned and requires only relatively low data rates.

The DVB decoder (integrated receiver decoder [IRD]) replaces a data modem in that the digital signals can be demodulated and decoded in order to be displayed. The decoder also edits and evaluates the return-channel data and additional information embedded in the DVB data stream which can, for example, then be passed on to the display for the creation of a graphic user interface.

Apart from the classical television receiver, the PC or "desktop video workstation" will be used in future as an actual terminal. This will surely not replace the television set, but if digital television is to be used for business applications such as stock exchange and banking it will be necessary to have a keyboard and a mouse for interactive direct communication.

Cassettes and video discs as well as the hard disc of a desktop video workstation could serve as storage media for digital television. The format of the digital video cassette (DVC) has already been determined within the framework of an international development project. Video disc formats are available under the description CD-video or DVD which enable the playback and/or recording of compressed television signals.

In the field of digital television expressions like "multimedia", "interactivity" and "data highway" play a specific role. At the moment no prognosis can be made of the progress of development in the technology which is concealed behind such expressions. The scenario shown in figure 1.1 presumably gives an adequate portrayal of the user environment in the era of interactive multimedia. Even at the present time it is certain that the technology of source coding and transmission to be described in this book will form part of this scenario. It will be supplemented by standards describing the broadband return channel, by concepts for so-called servers, in other words, for equipment which can provide and handle a great amount of data retrieved by a great number of participants, by standards for the software environment in the receiving devices (multimedia home platform [MHP]) as well as a myriad of communication protocols.

1.2 The Genesis of Recommendations for Digital Television

The evolution of the first concepts for digital television can only be understood in relation to the background of the worldwide development in the field of television engineering in the second half of the 1980s. While in Japan the multiple subsampling encoding (MUSE) system and in Europe the HD-MAC system, developed within the Eureka 95 project, were ready for the transmission of HDTV via satellite and through cable networks, the broadcasting community in North America felt threatened by the developments, as at that particular time they had nothing to offer to compete. Most particularly the distribution of HDTV over satellite envisaged in Europe and Japan would have endangered the traditional commercial enterprise of the broadcasters in the USA if there, too, satellite broadcasting had become the primary means of HDTV transmission. The stability of this field of enterprise is a result of the large networks acquiring advertising spots with a nationwide impact, whereas the actual transmission of the programmes within the network is left to the regional partners, so-called affiliates, who in turn secure their income through the sale of commercials which are of local and regional importance. It is clear that, in such a system, satellite distribution which cannot be regionalised would prove a destabilising factor. The growth of the cable business, which particularly in the USA resulted in new programmes being offered exclusively over cable, had already drastically changed the broadcasting scene. Therefore it was obvious that North America would start an initiative to develop terrestrial, and thus regionally transmittable, HDTV as an alternative to MUSE and HD-MAC.

The second initiative for the development of a terrestrially transmittable HDTV standard, but this time right from the beginning with the aim of achieving the breakthrough to digital television, was started in Scandinavia [APPELQ]. Looking back, it is difficult to identify the actual motivation for the start of the project HD-DIVINE. It was probably a blend of the conviction that HD-MAC was not technically viable, the wish of the national public broadcasting corporations to prevent the Scandinavian governments from granting terrestrial frequencies to commercial broadcasters by proclaiming a great Scandinavian awakening, and a considerable measure of pioneering spirit. Back in 1992 HD-DIVINE was able to demonstrate their first still incomplete system in public.

At the end of 1991 the first discussions in a close circle of Germans took place with the aim of taking stock of the worldwide technical situation of television and to settle the question as to which real development alternatives were available in Europe. Out of these first discussions the European Launching Group was born in the spring of 1992. This was a group with participants from all sections of the trade which met unofficially at the beginning and only in September 1993 evolved into the European DVB Project [REIMERS 1].

In Japan there was no official work on digital television for a number of years. The success of HDTV with MUSE as a transmission standard was not to be endangered. It was only in the summer of 1994 that the Ministry of Posts and Telecommunication (MPT) founded a Digital Broadcasting Development Office, which is responsible for the co-ordination of all achievements in development. A project structure, now called Advanced Radio Industries and Businesses (ARIB), supports this endeavour.

1.2.1 Work in the United States of America

The replacement of traditional analogue television by digital television was first thought of in 1990. It was recognised as the answer to a national initiative of the Federal Communication Commission (FCC) during 1987 with the objective to develop an HDTV standard in the USA to enable terrestrial and therefore locally confined transmissions of HDTV. At the initial stage, the "call for proposals" produced a veritable "gold rush" climate which led to 21 possible systems being submitted, some of which only tried to achieve the compatible improvement of NTSC and in this respect showed more similarity to PALplus than to HDTV. Many of the proposals promised HDTV quality but were nevertheless analogue standards in the above-mentioned sense and this was particularly true of Narrow-MUSE, a development for the USA by the national Japanese broadcasting authority, NHK [FLAHERTY].

By 1990 the list of the remaining system concepts which could be taken seriously had been reduced to nine. Two of them, both analogue systems, were conceived for parallel transmission of HDTV and NTSC (Simulcast). On 1st July 1990, General Instrument was the first company to recommend a fully digital concept. In 1991 there was only a total of five systems left in the race, all of which were intended for the transmission of HDTV. Four of them were digital and one (Narrow-MUSE) was analogue. The results of intensive testing on hardware prototypes between 1991 and 1992 reduced the list still further. Specifically Narrow-MUSE was eliminated [FCC].

In May 1993 the remaining system developers (the companies AT&T/ Zenith, General Instrument, DSRC/Thomson/Philips, and the MIT) stated their readiness to work together on the development of a single proposal, which since then has been known as the Grand Alliance HDTV System. This is still a system for terrestrial distribution of HDTV images. Multiple endeavours to initiate the development of transmission processes for digital television with something like the current image quality (standard-definition television [SDTV]) were resisted for a long time by the FCC. It was just before finalisation of the specification phase in the summer of 1995 that this quality level was integrated into the system.

The Grand Alliance HDTV System which was later adopted by ATSC, the Advanced Television Systems Committee, can be described in a somewhat simplified manner as follows [HOPKINS]: the source coding of the image signal is performed in accordance with the MPEG-2 standard as developed by the Moving Pictures Experts Group (Main Profile High Level – see section 4.3). This provision still holds good if SDTV images are to be encoded instead of HDTV images. The audio coding is in accordance with Dolby recommendation AC-3 and is multichannel-compatible. The forward error protection is multistage. The modulating process utilises vestigial-sideband (VSB) modulation of a single high-frequency carrier (see section 7.4). An 8-stage modulation (8-VSB) was specified for the terrestrial transmission and a 16-stage modulation (16-VSB) for distribution over cable.

Not all market partners in the USA were content with the system components chosen by the Grand Alliance. During the work of the Cable Television Laboratories a cable distribution standard using quadrature amplitude modulation (QAM) was developed which was very similar to the European system (see chapter 10). This standard now exists in addition to the VSB-based proposal by the Grand Alliance.

Broadcasters, unified under the National Association of Broadcasters (NAB), initiated a project for the investigation of the coded orthogonal frequency division multiplex (COFDM) process (see chapter 11), which was imported from Europe, as an alternative to the single-carrier transmission proposed by the Grand Alliance.

Finally, in the summer of 1994, the USA saw the first transmissions from a multiprogramme television satellite. Named DirecTV/USSB/DSS, it uses technology which is very similar to that developed in Europe (see chapter 9) by partner companies of the DVB Project and is practically identical to the European standard [BEYERS].

1.2.2 Work in Europe

The European DVB Project is the focal point of the development of digital television in Europe. This project came into being from the experience that developments in the complex field of electronic media can only be successful when all the important organisations working in this field participate in such a development and when the commercial interests are allowed to carry the

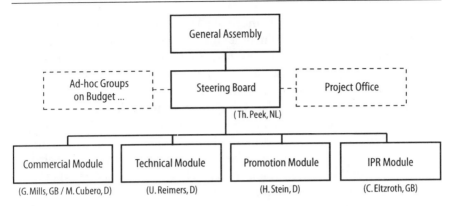

Fig. 1.2. Organisational chart of the European DVB Project

same weight as technical considerations in the definition of the technological objectives.

Figure 1.2 shows the organisational structure of the project. The number of member companies is 238, representing 25 European countries. Companies from the USA, Japan and Korea are represented by their European subsidiaries. Furthermore, associated members come from Canada and Australia. The members are either content providers, hardware manufacturers, network operators or regulatory bodies from the various countries. The Commission of the European Communities (CEC), the European Broadcasting Union (EBU), associations and standardisation institutes also have a special status for participation.

The commercial working group (Commercial Module) of the DVB Project is responsible for the definition of commercial requirements for the new systems from the viewpoint of the users in the fields of satellite, cable and terrestrial distribution as well as in the area of interactive services. These requirements form the basis for work within the Technical Module. After completion of the development the Commercial Module verifies the specifications for the new systems and, if necessary, passes them on to the Steering Board for the final passing of a resolution.

By means of a co-operation contract with the standardisation institutes ETSI and CENELEC an integration of specifications from the DVB Project into the regular standardisation procedures of both institutes is ensured.

At the time of the first work on DVB in Europe, the Moving Pictures Experts Group (MPEG), a working group reporting to the standardisation institutes ISO and IEC which are active worldwide, was already working on a procedure for the source coding of video and audio signals and had already started the design of the respective systems level (MPEG-2). The proposed system known as MPEG Audio for the source coding of audio signals, in mono and stereo, was already in the final phase of standardisation. The European DVB Project decided that in order for the technological solution used by the DVB Project to find a wide international basis, digital television should utilise the MPEG standard. This decision led to an intensive co-operation between numerous European organisations and MPEG with the result that the research conducted in some places in Europe – with a different approach to image coding – was put aside in favour of a worldwide standard. The DVB Project's strategic aim to support a single worldwide solution was partly responsible for the decision taken by the Grand Alliance in the USA to also select MPEG-2 for the source coding of image signals. Japan, too, is expected to adopt MPEG-2 for source coding.

The first important result of the DVB Project emerged in the second half of 1992 under the leadership of the author. This was the report to the European Launching Group [WGDTB] on the "Prospects for Digital Terrestrial Television", which was presented in November 1992. This report showed how, and with which aims, a DVB system for Europe could be developed. The report was comparatively heavily weighted in favour of terrestrial transmission and towards HDTV as a probable quality objective. In this respect it was a product of its time and took into account the fact that at the end of 1992 the official European development policy was still centred on the satellite transmission of HD-MAC.

Even during the production of the above-mentioned report the European DVB Project was able to profit enormously from the existence of research consortia which were sponsored by funds from the EC (RACE dTTB, SPECTRE, STERNE) or from the German Federal Ministry for Research and Technology (HDTV-T), or which were organised by private enterprise (HD-DIVINE). During the further development of the work the said consortia carried on playing an important role, particularly in the definition of the standards for the terrestrial transmission of DVB.

The first complete system specification was the recommendation for satellite transmission adopted by the Technical Module of the DVB Project in November 1993 [REIMERS 3, REIMERS 4]. In December the steering board agreed with this recommendation and in November 1994, by the unanimous decision of all member states of ETSI, this became the European Telecommunication Standard ETS 300 421. In January 1994 the specification for DVB distribution over cable (ETS 300 429) followed and since then numerous other specifications have been developed and adopted.

In the following chapters many of the results of the work of the DVB Project are described in detail. The contribution of the authors in the development of these results will be incorporated in the text. The complexity of the DVB scenario, however, does not permit the inclusion of all the DVB results, such as the specifications of technologies to be used in interactive channels etc.

1.2.3 Work in Japan, Canada and Korea

As already mentioned, in Japan the work on the development of digital television officially commenced in the summer of 1994. However, unofficially this topic had already been examined in many places since a lot of European subsidiaries of Japanese equipment manufacturers were already members of the DVB Project. Therefore early reports from Japan (personal communication) showed a close correspondence between Japanese and European conceptions. As in Europe, MPEG-2 was chosen for the source coding and for the system level. Quadrature phase shift keying (QPSK) was also adopted as the modulation procedure for the satellite transmission [NHK]. Quadrature amplitude modulation (QAM) is used for the distribution in cable networks. COFDM is being considered for the terrestrial transmission. Recent talks with Japanese organisations in the field of work of the International Telecommunication Union (ITU) have shown that the European standards may even be adopted for forward error protection.

For many years the group Advanced Broadcasting Systems of Canada (ABSOC) had been working on the development of recommendations for digital television. ABSOC, although not interested in an independent standard, owing to the geographic and commercial proximity of Canada to the USA, saw itself as an organisation which would still like to make its own contribution to the North American development of digital television. Therefore ABSOC co-operated closely with the European DVB Project and tried to transplant many European solutions into the work of the Grand Alliance (for example, service information – see section 5.4).

In 1993 a Korean DVB Project was founded with the participation of organisations from South Korea. In view of the fact that the most important South Korean companies in entertainment electronics already co-operate through their European subsidiaries in the DVB Project, the Korean group has not developed any important autonomous system recommendations.

1.3 Objectives in the Development of Digital Television

One of the most important questions asked with regard to the development of DVB is about its commercial reason. Why should a system for entertainment television be commercially successful, in a saturated market, in competition with a multitude of existing television standards, some of which have been in use for many years and therefore have a great number of compatible receivers?

At the beginning the DVB Project compiled a catalogue of possible goals, which could be described as classical or typical of broadcasting [REIMERS 2]:

(1) Digital television might enable the transmission of very high-quality HDTV images, possibly even via future terrestrial broadcasting networks.

(2) DVB might enable the broadcasting of programmes of contemporary technical quality using narrowband channels for transmission, or it might enable an increase in the programme range within existing transmission channels.

(3) DVB might be the method of broadcasting to low-cost pocket TV receivers, equipped with built-in receiving antennas or short rod antennas, which would guarantee stable reception for a number of television programmes.

(4) Television receivers in vehicles (trains, busses or cars) might be served by DVB with broadcasts of a similarly superb quality as that at present transmitted to radio receivers by Digital Audio Broadcasting (DAB), i.e. DVB might enable stable reception in moving vehicles even over difficult radio channels and at high speeds.

(5) Moreover, as a data transmission technique, DVB might retain the typical characteristics of digital technology, such as the stability of the transmission within a very clearly defined coverage area, the possibility of simple transmission over telecommunication lines as one service among many, and the possible integration into the world of personal computers.

As the work on DVB progressed the objectives changed considerably. HDTV has not been completely lost from view, but it has lost its role of primary objective. The servicing of portable receivers has remained an objective during the development of standards for terrestrial transmission, but it is not as important as originally envisaged. From the extensive list of optional parameters for a terrestrial standard (see chapter 11) it is possible to choose operational modes suitable for portable reception. Finally, mobile reception was not a part of the original user requirements of the DVB system, although this requirement is currently evolving in some countries. The DVB system specification for terrestrial transmission has, in the meantime, demonstrated its capability to provide for stable mobile reception up to very high speeds.

In the course of time "data container" has become the key concept with regard to the definition of the objectives of DVB [REIMERS 5]. This concept illustrates the idea which underlies the design of the transmission standards for all methods of transmission. A data container is defined by the fact that a maximum amount of data per unit of time can be transmitted in it quasi error free (QEF). It does not matter what kinds of data are transmitted as long as they are packetised and supplemented with additional data, such as synchronous information, in accordance with the rules of the various DVB standards. With this background, the question about the reason for DVB can now be answered as follows:

(1) DVB enables a multiplication of the number of television programmes which can be broadcast on one transmission channel or in one data container.

(2) DVB makes the broadcasting of radio programmes possible and enables data transmission for entertainment and business purposes.

(3) DVB makes a flexible choice of image and audio quality possible, including the choice of HDTV as long as the resulting data rate does not exceed the capacity of the data container.

(4) For use in connection with pay services there are very secure coding methods which ensure that unauthorised access to such services is extremely difficult, if not impossible.

(5) Furthermore, as a data transmission technique DVB incorporates typical characteristics for the utilisation of digital technology, such as the stability of the transmission within a clearly defined coverage area, the possibility of simple transmission over telecommunication lines, as one service among many, and the possible integration into the world of personal computers.

1.4 Data Reduction as the Key to Success

The data containers already mentioned in section 1.3 are available in various capacities, in accordance with the transmission medium. Specifically, for example, in satellite transmission and distribution on cable networks, the capacity is in the range of 40 Mbit/s (see chapters 9 and 10), while for terrestrial transmission the capacity could be 20 Mbit/s (see chapter 11).

Regarding the digitisation of colour television signals in compliance with ITU-R Recommendation BT.601 [ITU 601], it is obvious that without an efficient reduction in the data rate not even a single television programme, let alone many, can be transmitted within a data container. Recommendation 601 calls for a data rate of 216 Mbit/s for a complete television image, including blanking intervals, or 166 Mbit/s when blanking intervals are buffered and can be ignored (see chapter 2). Hence data-rate reduction is indispensable (see chapter 4).

Similar considerations apply to audio signals. The original data rate for stereo signals in CD quality is 1.4 Mbit/s (see chapter 2). Therefore they would require an exorbitantly large part of the capacity of a data container, unless here, too, efficient data-rate reduction is achieved (see chapter 3).

In figure 1.3 there is an illustration of data-rate reduction for video signals. There are three groups of input data rates on the ordinates, values which are produced in television studios before data-rate reduction. In each case these are net values, data rates describing the video signals without blanking inter-

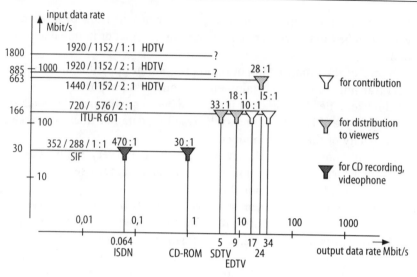

Fig. 1.3. Examples of compression ratios for video data reduction

vals. Current studio technology uses 166 Mbit/s (see above). For HDTV, values between 663 Mbit/s and 1800 Mbit/s are being discussed. If, right from the start, systems are to be produced with a greatly reduced video quality (which would be roughly comparable to a medium-quality VHS recording), it would seem that the utilisation of original video material with a reduced video quality would suffice (MPEG-1 source input format – SIF, see chapter 4). The luminance quota of SIF images is only 288 lines, with 352 pixels per line, and it is progressive, which means that an image is represented by 25 frames per second. In contrast to what is required in today's studio standards, the chrominance components are added to the luminance signals line-sequentially at a ratio of 4 : 2 : 0. The resulting data rate of quantised source signals of 8 bits per pixel is therefore 30.4 Mbit/s.

The funnels in figure 1.3 symbolise the data-rate compression by means of the source coding techniques introduced in chapter 4. Significant examples of output data rates after completion of compression are indicated and explained. "ISDN" stands for the compression by means of ITU-T Recommendations H.261 or H.263 [ITU H.261, ITU H.263], which also aim for videophone applications with the ISDN data rate of 64 kbit/s. "CD-ROM" symbolises the use of data reduction to produce video signals for the storage on CD formats such as CD-Interactive or CD-ROM. The MPEG-1 standard is used as the compression method. "SDTV" (standard-definition television) refers to a video quality which does not warrant an error-free reproduction, but the sum of whose artefacts in most of its picture content is comparable to the present PAL television. It seems that 4–6 Mbit/s are sufficient for SDTV when, as with all higher data rates, one of the standards known under the generic term of MPEG-2 is employed. "EDTV" (enhanced-definition television) stands for virtually transparent video transmission, at a quality capable of being produced in a studio in accordance with ITU-R Recommendation BT.601.

1.5 Possible Means of Transmission for Digital Television

Most people, when thinking of the technical means of transmission for television programmes, probably think first of terrestrial transmission, which is wireless broadcasting of programmes from antenna masts standing on the ground.

However, already during the phase of planning the introduction of Digital Audio Broadcasting (DAB), the provision of broadcasting frequencies for new terrestrial services – at least in Europe – proved to be exceptionally problematic. The same is also true for television, because it is neither possible, at short notice, to clear channels now in use for analogue transmissions for the exclusive use of digital television, nor are there unused frequency resources which could be allocated for television broadcasting. In this respect the terrestrial transmission of DVB is dependent on finding new technical concepts for its broadcasting standard, enabling the utilisation of the few frequencies available to reach a greater number of members of the public. A long-term strategy has to be found for the reallocation of frequency resources. Another option would be to introduce digital television in areas where spectrum capacity can be found, but this would result in "geographical islands" being created.

At present in the United States of America there are approximately 1700 television transmitters in use [JANSKY]. This represents such a minimal utilisation of the spectrum capacity designated for television broadcasting that the work of the Grand Alliance targets the introduction of DVB into existing gaps of the frequency spectrum, the so-called "taboo channels" [JANSKY]. These taboo channels could, in principle, be used for the transmission of analogue NTSC signals but, to date, are not in use because they cannot be located in proximity to a channel actually in use, or are located where there might be mutual interference with a local oscillator frequency. The transmission system developed by the Grand Alliance should now be in a position to allow broadcasts in some of these channels which are at present unused. This means that in a comparable service area the interference effect of a DVB transmitter on an existing NTSC receiver must be reduced, for example by reducing the broadcasting power in comparison with an NTSC transmitter, but at the same time the robustness of the digital system must be sufficient to enable interference-free DVB reception despite the presence of the NTSC transmitter.

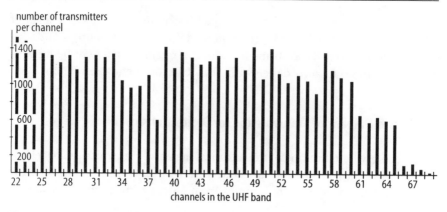

Fig. 1.4. Occupancy of the UHF band for television transmitters in Europe

In comparison with the USA the situation in many parts of Europe is distinctly more complex. It is typical for the national broadcasting services in European countries to have taken on the task of total coverage for the general public. This means that there are a great number of transmitters with varying power levels. Figure 1.4 shows the European utilisation of the UHF band (channels 22 to 69) for television transmitters of all power levels [WGDTB]. It is easy to see that in channel 22 alone there are almost as many transmitters as in the whole of the United States of America. At first sight it looks as if the only reserves are within the range of channels 34 to 38 and above channel 60. In actual fact, different allocations in both these ranges exist in some European countries. Most particularly the channels above 60 are still being used in many countries for military services.

A White Paper had been in existence in the United Kingdom since the summer of 1995 in which the general conditions for the introduction of terrestrial digital television and also for terrestrial DAB were defined. Concerning the terrestrial transmission of DVB signals, this document showed a way in which a major percentage of the public could be offered as many as 28 terrestrially transmitted programmes, distributed over 6 channels. However, contrary to the original plan, these channels have been mainly allocated using conventional frequency planning. The concept of the so-called single-frequency networks (see chapter 11) has therefore lost significance in the United Kingdom. In November 1998 DVB-T services were successfully introduced there.

Scandinavia is a region in Europe which, due to its relative border location, has limited interference from the transmission networks from neighbouring countries, and moreover the number of terrestrially transmitted analogue programmes is so limited that there are frequency reserves within the whole UHF band which could be used for the introduction of DVB. In Sweden, in particular, such utilisation seems already to have been decided.

In Europe a recommendation for a new distribution of the available frequency spectrum was developed by a specialist group from the European Radiocommunications Office [ERO] on the basis of an analysis of the present allocation of frequency bands to the various types of users. This recommendation, which was presented in the autumn of 1995 during the European Conference of Postal and Telecommunications Administrations [CEPT], includes the proposal that, by the year 2008, in all those regions where the channels above channel 61 are not yet utilised for radio broadcasting, these channels ought to be made available for DVB. By the year 2020 large parts of the UHF spectrum should then be utilised by DVB, while at the same time the number of transmitters for analogue standards should be reduced.

In Chester, in the summer of 1997, more than 30 European countries signed a multilateral co-ordination agreement which included all the necessary rules and technical parameters that were to enable the start of frequency planning for terrestrial digital television in all of Europe.

In the summer of 1998 the German federal government approved a document on the future of broadcasting in Germany which was proposed by a grouping called "Initiative Digital Broadcasting". This document proposes that digital terrestrial television should start to be introduced by the year 2000 and that the last analogue TV transmitter should be switched off by 2010.

The possibilities for the introduction of DVB by satellite and cable are much less affected by the necessity of long-term planning than those for terrestrial transmission. Therefore, the mainstay of the introduction strategy for Europe is satellite distribution. The initial stages in this strategy go back to the spring of 1996. The ASTRA 1E satellite, in its orbit position of 19.2° east, was ready for DVB transmissions since October 1995. Figure 1.5 shows a possible scenario for the distribution of DVB with the satellite taking the central position. Following the multiplexing of individual compressed audio, video and auxiliary data signals into programmes 1 to n, these are combined to form a single data stream (the data container). This is then transferred to the satellite uplink after suitable processing. The satellite then broadcasts the data container to the direct-to-home (DTH) receiver, to cable head ends and to the locations of terrestrial transmitters. The signals are received in the cable head ends and demodulated and, possibly not completely, decoded. Next, if required, programmes are remultiplexed within a container and then further distributed within the cable network by means of a specific modulation technique. Of course, cable programmes can also be distributed to the cable head ends by other means, for example, via telecommunication links. The procedure for cable networks is also applicable to community antenna systems. In many countries in Europe, and particularly in Germany, the decision to introduce DVB into cable networks has already been made. At the sites with terres-

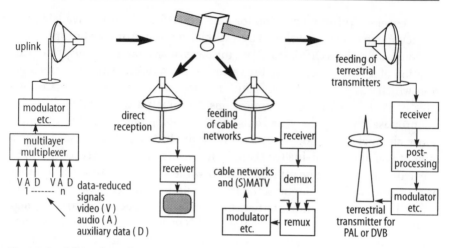

Fig. 1.5. Possibilities of a satellite-based distribution of DVB

trial transmitters for analogue or digital television, satellite signals can also be received, decoded and demodulated. They can then be converted to any other transmission standard, such as PAL or SECAM, before transmission to the public [ROY].

When, to which extent, and in which European countries digital television will be brought to the viewers by telecommunication lines is less a technical question than a commercial one, or rather a question of communication policy. In Great Britain, where British Telecom does not have the right to broadcast and where broadcasting is regarded as the simultaneous distribution of a programme to as few as two recipients, there may be a great incentive to utilise the copper-wire and optical-fibre technology which British Telecom owns for the transmission of programmes to telephone customers. In Germany, on the other hand, Deutsche Telekom AG (German Telecom) has the monopoly not only of telephone networks but it also operates all of the higher layers of the cable networks. It is therefore highly unlikely that Deutsche Telekom will introduce DVB services both on cable and via telephone lines at the same time or in the same regions of the country.

1.6 Standards and Norms in the World of Digital Television

The development of digital television has led to technical specifications for a myriad of system components, ranging from the description of coding algorithms to that of modulation procedures, transmission parameters, hardware components, device interfaces, or techniques for interactive channels. Many organisations, such as MPEG, the DVB Project and the Digital Audio Visual

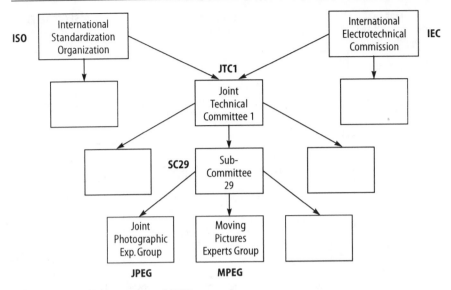

Fig. 1.6. Organisational integration of MPEG

Council (DAVIC), participate in the development of specifications. The transformation of the specifications into standards, however, does not lie within the competence of these groups but is the responsibility of European or even world standardisation institutes. The International Standardization Organization (ISO), the International Electrotechnical Commission (IEC), the European Telecommunications Standards Institute (ETSI) and the Comité Européen de Normalisation Électrotechnique (CENELEC) all play an important role in the DVB context. The task of creating unity in the system concepts and standards for DVB which are developing the world over is undertaken by the International Telecommunication Union (ITU).

The integration of the Moving Pictures Experts Group (MPEG) in the work of the ISO and IEC organisations is shown in figure 1.6. The ISO/IEC Standards (IS 11 172, IS 13 818 [Parts 1,2,3]) are a result of this integration of MPEG.

The DVB Project has added a long list of specifications to these ISO/IEC standards. All the work packages listed in figure 1.7 have been completed.

The DVB Project is also working closely with groups outside the actual field of digital television: in connection with the future development of digital video recorders (DVC) and digital multimedia discs (DVD) as well as with respect to work being done by the telecommunication authorities in the field of switching and high data-rate network (trunk) technologies.

A co-operation contract has been agreed between the DVB Project and the already-mentioned organisations ETSI and CENELEC to the effect that the specifications originating from the Project are passed on to one of these or-

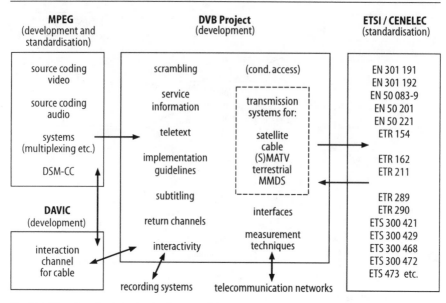

Fig. 1.7. Allocation of tasks in the development of digital television

ganisations. The JTC Broadcast (joint technical committee) decides to which standardisation institute each specification will be delivered; as a general rule one can say that those describing a system will be submitted to ETSI and those describing a piece of equipment or its interfaces will be submitted to CENE-LEC. Up to the present, more than twenty European Telecommunication Standards (ETS) have been agreed with ETSI, and three further European Norms (EN), with CENELEC. Eleven ETSI Technical Reports (ETR) concerning DVB technology have also been published by ETSI.

2 Digitisation and Representation of Audio and Video Signals

The sources of the audio and video signals, such as microphone and camera, produce analogue signals, i.e. signals of an unlimited range of values, progressing continuously in the time domain. For digital processing the signals must be discretely sampled at regular intervals, quantised and coded. These processes are subsumed under the term 'digitisation' and are performed by analogue-to-digital converters (ADCs). Reconversion into analogue signals is carried out by digital-to-analogue converters (DACs).

Between these two converters the signals are transmitted in bit-parallel or bit-serial form and further processed. Certain signals and/or periods of time within periodically recurring signal patterns (frames) can be used specifically for synchronisation and for the transmission of auxiliary information. The signal shapes and data rates resulting from this procedure are here classed under the term 'signal representation'.

2.1 Sampling and Quantising

The frequency of the sampling clock obeys Nyquist's criterion which postulates that the sampling rate of the analogue signal must be at least twice the highest frequency to be sampled, so as to avoid aliasing in the digitised signal [SCHÖNFD 1]. The analogue signal must be prefiltered accordingly. Usually a sampling frequency higher than the Nyquist limit (oversampling) is chosen to increase the distance between the baseband spectrum and the repeat spectra and so reduce the implementation requirements for the filter.

The number of steps required for the quantisation and coding of the analogue signal is determined by the amplitude resolution of the digitisation process. This entails 2^b quantisation steps of the same size for linear, binary coding with b bits. The attainable level range is limited at the lower end by the quantisation error of one step and at the upper end by the limiting effect of the ADC which sets in as soon as the quantisation characteristic of the latter is overdriven. This level range is also referred to as system dynamics.

Under very general conditions relative to a stochastic analogue signal to be quantised, the power of the inherent quantisation error can assume the value of $Q^2/12$ if Q represents the size of one of the quantisation steps. If the chosen step Q is small enough, this quantisation error can be conceived as a uniformly distributed random sequence with zero mean value and with a white spectrum which is not correlated with the original signal [SCHÜSSLER]. This may also be called quantisation noise and can be used to define a signal-to-noise ratio. With deterministic signals, such as those often found in video applications, these requirements are not satisfied in every case, owing to which the quantisation error can no longer be described in such a general way. Even so, the above assumption offers a sufficient estimation of the disturbing effect of the quantisation error.

The following is valid for video signals when only the range of the picture signal S_V is digitised:

$$S_V / N_Q = b\,(6\ \text{dB}) + 10{,}8\ \text{dB}\ , \tag{2.1}$$

for audio signals, when referred to the r.m.s. value of S_A:

$$S_A / N_Q = b\,(6\ \text{dB}) + 1{,}8\ \text{dB}\ . \tag{2.2}$$

The different constants in both formulae can be derived from S_V (peak-to-peak value of the signal amplitude) and S_A (r.m.s. value of the signal amplitude) [SCHÖNFD 1]. The equivalent r.m.s. value of the quantisation error in both cases is $N_Q = Q / \sqrt{12}$ in the sense already explained.

The quantisation error can also depend on the frequency of the analogue signal. The layout of the sampler in the ADC (transitional behaviour, aperture jitter) is important here.

2.2 Digitising Video Signals

The standardised sampling rate of 13.5 MHz for the R, G, B or Y baseband signals [ITU 601], together with the quantisation of 8 bits per sample specified for broadcast transmission, leads to a data rate of $H_0 = 108$ Mbit/s for each of these signals. For each of the chrominance signals C_B and C_R the sampling rate is reduced to 6.75 MHz and the data rate to 54 Mbit/s. The Y sampling rate of 13.5 MHz is defined as an integral multiple of the line frequency for the 625-line standard as well as for the 525-line standard. This results in a duration of the 'active' line of 720 samples in both standards. Furthermore, this choice facilitates the design of the pre-filter required in front of the ADC, as already mentioned. The same applies to the chrominance components.

For an *RGB* transmission (format 4 : 4 : 4) this results in a total data rate of $H_{0\,Total} = 324$ Mbit/s or, for a $Y/C_B C_R$ transmission, due to the reduced sam-

Table 2.1. Digitisation characteristics and data rates of video signals in accordance with [ITU 601]

Signals	Clock [MHz]	b [bit]	H_o [Mbit/s]	H_{oTotal} [Mbit/s]	Format
R	13.5	8	108		$4:4:4$
G	13.5	8	108		ITU 601
B	13.5	8	108	324	
Y	13.5	8	108		$4:2:2$
C_B	6.75	8	54		ITU 601
C_R	6.75	8	54	216	

Table 2.2. Pre-filter specifications in accordance with [ITU 601]

	Y/RGB Signals	$C_B C_R$ Signals
Pass band: Insertion loss	ascending from ±0.01 dB at 1 kHz to ±0.025 dB at 1 MHz ±0.025 dB/1 ... 5.5 MHz; ±0.05 dB/5.5 ... 5.75 MHz	ascending from ±0.01 dB at 1 kHz to ±0.05 dB at 1 MHz ±0.05 dB/1 ... 2.75 MHz
Group delay time	ascending from ±1 ns at 1 kHz to ±3 ns at 5.75 MHz	ascending from ±2 ns at 1 kHz to ±6 ns at 2.75 MHz ±12 ns/2.75 ... 3.1 MHz
Stop band: Insertion loss	≥12 dB/6.75 ... 8 MHz ≥40 dB/8 ... 13.5 MHz	≥6 dB/3.375 ... 4 MHz; ≥40 dB/4 ... 6.75 MHz

pling rate of 6.75 MHz for the chrominance components (format $4:2:2$), a rate of $H_{oTotal} = 216$ Mbit/s. The results for both formats are compiled in table 2.1.

A quantisation of 10 bits per sample is also possible as an option for signal transmission and processing in a studio environment and within equipment, in which case the data rates will increase accordingly. Of late, a sampling rate of 18 MHz, instead of 13.5 MHz, has been suggested for an aspect ratio of $16:9$, in order to obtain a horizontal resolution which is adapted to this picture format [ITU 601]. In this way the data rates stated increase by the factor of 4/3.

The specifications for filters before of the ADCs for the sampling rates of 13.5 or 6.75 MHz are shown in table 2.2.

Filters after the DACs generally have lower requirements to fulfil as most receivers in the frequency range under consideration already have a low-pass characteristic. For equipment with subsequent analogue signal processing the filter characteristics have to be designed accordingly.

Due to the sample-and-hold function of the DAC the output signal has a spectrum which decreases with higher frequencies in accordance with an si-

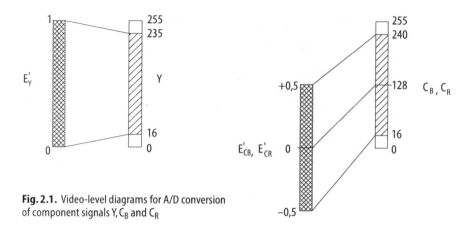

Fig. 2.1. Video-level diagrams for A/D conversion of component signals Y, C_B and C_R

function [SCHÖNFD 1]. The compensation of this frequency response can take place passively in the post-filter or actively by means of an equaliser.

The level ranges for the conversion of analogue component signals into the digital $Y/C_B C_R$ component signals are smaller than the system dynamics available for the digitisation as shown in figure 2.1. The values 255 and 0 are reserved solely for the coding of the synchronising signals. In addition, at the upper and lower ends of the level range some quantising steps are reserved as a protection against overdriving. Generally, the amplitudes are converted to the binary code, with the luminance signal Y not being offset and the bipolar chrominance signals C_B and C_R being offset by their addition to the centre value of the system dynamics.

The correct control of the level ranges requires fixed signal values for 'black' or 'uncoloured'. This is effected by means of appropriate circuits directly in front of the ADC. Depending on the particular requirements, these can be simple clamping circuits or more complex control circuits to retain the black or uncoloured value with changing signal averages such as have been used for many years in television engineering. The feedback of digitised reference values into the aforementioned control circuits can also be used to compensate for tolerances in analogue signal processing, including that of the ADC [IRWIN].

2.2.1 ADCs and DACs for Video Signals

For video ADCs, almost the only technique still used nowadays is the parallel method or flash conversion. The block diagram of such a type of converter is shown in figure 2.2 [SCHÖNFD 1, TIETZE]. The analogue input signal $u_1(t)$ is fed to a chain of 2^b-1 comparators which receive their references from the taps of a voltage divider that divides the reference voltage U_{REF} into 2^b-1 parts. The comparators are activated by the sampling clock and in this way

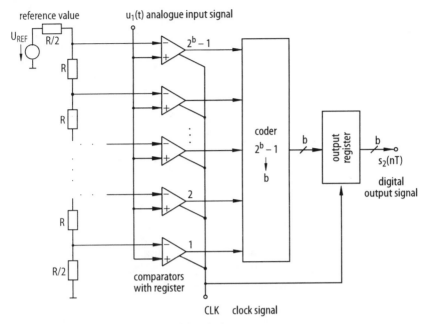

Fig. 2.2. ADC in accordance with the parallel method

function as samplers and quantisers at the same time. Their $2^b{-}1$ output signals represent the digital value in the form of a thermometer scale, the digital value is converted by a subsequent coder into one of the usual binary codes with b bits in the digital output signal $s_2(nT)$.

The hold function for the sampling value is performed by registers at the outputs of the comparators (not shown in figure 2.2) as well as by the output register of the ADC.

Each of the b bits of the input signal in video DACs controls a current source which, according to the weight of the respective bit, contributes to the sum current at the output. Thus the magnitude of this sum current varies in accordance with the value of the digital input signal.

In the circuit diagram of figure 2.3 the individual currents which are allocated to the bits of the digital input signal $s_1(nT)$ are graded by powers of two with the help of an R-2R resistance network. The switches $b_7 \ldots b_0$ connect the currents either to the analogue output $i_2(t)$ or to the ground, depending on their bit pattern [SCHÖNFD 1]. The correct function of the device depends mainly on the simultaneous current switching of the circuit; a temporal misalignment of individual currents can lead to correspondingly large transients ('glitches') in the sum current $i_2(t)$ and therefore to take-over errors in the analogue output signal $u_2(t)$.

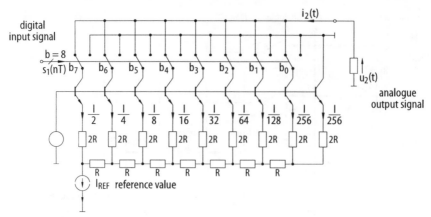

Fig. 2.3. DAC with an R-2R resistance network

Other models of video DACs use equal individual currents which, when added, are weighted by powers of two by an R-2R resistance network [TIETZE]. Hybrid forms between this variant and the one depicted in figure 2.3 are also known.

2.2.2 Representation of Video Signals

Figure 2.4 [ITU 656] shows the temporal relationships between analogue and digital video signals and the multiplex schedule of the video signals. The blanking of the digital signals begins prior to the rising-edge centre of the analogue synchronising signal, the duration of this blanking (144 samples for the 625-line standard, 138 samples for the 525-line standard) being shorter than that of the analogue signal. The fact that the sampling frequency and the duration of the 'active' lines for both standards are the same warrants that the temporal relationships between analogue and digital signals are almost independent of the standard.

At the beginning and at the end of the 'active' line the digital signal contains synchronising words (video timing reference signals), referred to as 'start of active video' (SAV) and 'end of active video' (EAV). Each of these words requires four time slots of the $Y/C_B C_R$ multiplex. Included is information about the actual field, the vertical blanking interval, and the beginning (EAV) or end (SAV) of the blanking. In addition, four further bits are coded such that they can correct one bit error and detect two bit errors in this information at the receiving end.

The multiplex schedule, as depicted in the last line of figure 2.4, applies to the parallel transmission of the eight or ten data streams, depending on the bit values. Apart from this, an additional ninth or eleventh line for the multiplex clock (27 MHz) is included so that no clock recovery at the receiving end

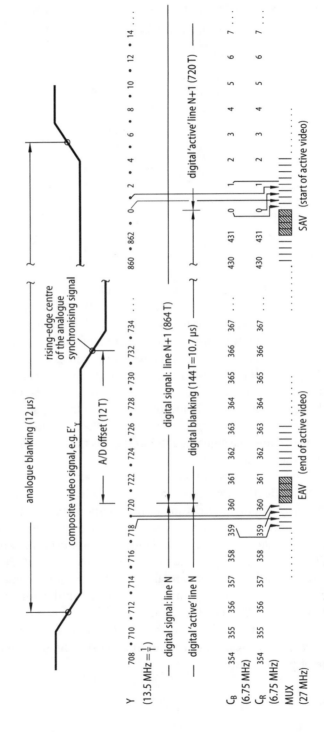

Fig. 2.4. Temporal relations between analogue and digital video signals (625-line standard) as well as multiplexing schedule of the signal components

Table 2.3. Digitisation characteristics and data rates of video signals with irrelevance reduction

Signals	Clock [MHz]	Values/ [line]	Lines	H_o [Mbit/s]	H_{oTotal} [Mbit/s]	Format
R	13.5	864	625	108		$4:4:4$
G	13.5	864	625	108		ITU 601
B	13.5	864	625	108	324	
Y	13.5	864	625	108		$4:2:2$
C_B	6.75	432	625	54		ITU 601
C_R	6.75	432	625	54	216	
Y	13.5	720	576	83		$4:2:2$
C_B	6.75	360	576	41.5		only active image
C_R	6.75	360	576	41.5	166	
Y	13.5	720	576	83		$4:2:0$
$C_B C_R$	6.75	360	576	41.5	124.5	only active image
Y	6.75	360	288	20.7		$4:2:0$, SIF
$C_B C_R$	3.375	180	288	10.4	31.1	only active image

is necessary. Directly after the end of the blanking (SAV), data words with the signal components C_B, Y, C_R, Y are transmitted in the succession here indicated. This sequence repeats itself until the beginning of the next blanking (EAV) and in this way ensures the allocation of two luminance signal values to the corresponding two chrominance signal values.

For the serial transmission (digital serial components, DSC) an additional coding, after the parallel-to-serial conversion of the data stream, is required with which, apart from the spectral shaping of the signal, support of the clock recovery and the word synchronisation at the receiving end is achieved. The bit rate of the serial signal is 270 Mbit/s; it is mainly used for signal distribution in the studio [ITU 656].

Possible data rates of video signals are listed in table 2.3, taking into account the effect of simple methods of data reduction. Irrelevance reduction is what this is about, i.e. omission of information which is not required for the relevant application. The frame frequency of the signal formats in table 2.3 is always 25 Hz and the word width 8 bits. By eliminating blanking intervals during transmission it is possible to reduce the total data rate H_{oTotal} from 216 to 166 Mbit/s, and by line-sequential C_B/C_R transmission, to 124.5 Mbit/s (format $4:2:0$). Finally, a subsampling by a factor of 2 in the horizontal and vertical directions results in a further reduction of the data rate H_{oTotal} by a factor of 4, to 31.1 Mbit/s (source input format, SIF). This signal format refers to a progressive picture scanning without line interlacing, independent of the later type of picture display.

Table 2.4. Digitisation characteristics and data rates for the transmission of audio signals

Standard	Clock [kHz]	H_{oMono} [kbit/s]	$H_{oStereo}$ [kbit/s]	Uses
DSR	32	512	1024	Digital satellite radio
CD	44.1	706	1412	Audio CD
AES/EBU	48	768	1536	Professional audio studio

2.3 Digitising Audio Signals

The sampling parameters and the data rates of the more usual digital audio systems are shown in table 2.4. Throughout, a quantisation of 16 bits per sample is assumed for the transmission.

The clock frequencies of 32 and 48 kHz are derived from multiples of the sampling frequency of telephone signals (8 kHz). The numerical ratio of the clock frequency of 44.1 kHz to the line frequencies of the TV systems for both 625-line and 525-line scanning is relatively simple. The reason for this lies in the fact that in order to produce moulds during the mastering of CDs the digital audio signals have to be stored temporarily for the playing time of the carrier (for approx. 70 minutes). It is often video recorders that are used for this purpose, therefore it is convenient to choose an audio clock frequency which is quasi-compatible with the line frequencies.

How the chosen quantisation affects the levels is illustrated in figure 2.5. The system dynamics, for a 16-bit and a 20-bit system, is given in (2.2). A

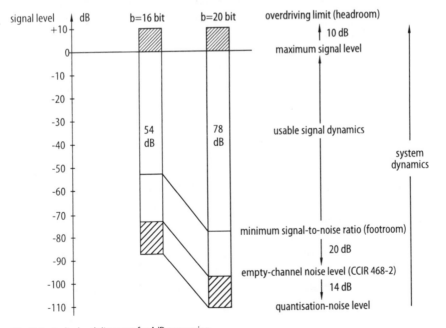

Fig. 2.5. Audio-level diagrams for A/D conversion

headroom of 10 dB is allowed for overdriving tolerance since the driving control for audio signals is not easy to handle and an abrupt limiting must definitely be avoided [ZANDER]. The aurally compensated evaluation of the quantisation noise (with some tolerance) leads to an empty-channel noise level of 14 dB above the quantisation noise level [JAKUBOW, HESSNM]. The remaining differences of 74 dB or 98 dB represent the actual usable level ranges. If an S/N ratio of 20 dB is allowed at the lowest passages, then a usable signal dynamics of 54 dB and 78 dB is obtained.

The level diagrams illustrate why a quantisation of 20 bits or more is appropriate in the professional studio. On the other hand, during transmission the effective quantisation can be reduced to between 14 and 12 bits by non-linear quantisation (companding) or by a dynamic adjustment of the level range to the quasi-instantaneous value of the signal without the transmission quality being, as a rule, audibly impaired [HESSENM].

2.3.1 Representation of Audio Signals

Apart from the bit-parallel processing of digital audio signals (which chiefly takes place inside equipment), bit-serial transmission has been adopted in the professional studio and for consumer use. One of the signal formats used (AES/EBU) is described in [IEC 958]. It is structured in subframes of 32-bit lengths, each of which, apart from one audio-signal sample (of a length of up to 24 bits), contains the synchronising characters, the parity bit and one information bit. Two of these subframes together form a frame and transport the dual-channel or stereo sound. A block is formed by 192 frames in which the channel status is transmitted by the sum of all information bits. Information about source and destination, about parameters, the allocation of the signals and, further, a time code and error-protection bits are included. The data rates of these bit-serial audio signals prior to modulation (biphase mark) are shown in table 2.5 and are dependent on the sampling frequencies of the source signals.

2.3.2 ADCs and DACs for Audio Signals

There are no guiding values for pre- and post-filters applied to audio signals, as recommended for video signals in accordance with table 2.2. The stop-band attenuation of these filters ought in principle to achieve the value of the signal-to-quantisation-noise ratio. In the case of a clock frequency of 48 kHz, a pre-filter cut-off frequency of 15 kHz and $b = 16$ bits, a Cauer low-pass filter of grade 9 is required to meet the above requirement. This very conveniently chosen example shows that this type of pre-filter – if necessary with phase equalisation – is costly to implement. This cost can, however, be con-

Table 2.5. Data rates for bit-serial audio signals

Clock [kHz]	b_{Max} [bit]	H_{oSer} [Mbit/s]
32	24	2.048
44.1	24	2.8224
48	24	3.072

siderably reduced by oversampling and digital signal processing, which will be discussed later. The same applies to the post-filter.

The weighting process is widely used in ADCs for audio signals (figure 2.6). By means of a register all b bits of the digital output signal are tentatively arranged in succession, commencing with the most significant bit (MSB). The result, after the D/A conversion, produces a comparison signal, which is fed to a comparator, the other input of which contains the sampled and retained analogue input signal. The greater-smaller decisions of the comparator determine the final value of the output signal. For this the clock frequency of the successive-approximation register must be at least b times higher than the sampling frequency of the input signal [TIETZE].

Progress in semiconductor technology allows the use of a considerably higher sampling frequency for the converter than would be required in accordance with the Nyquist theorem. An advantage of this oversampling is the separation of the frequency locations of the baseband spectrum and the sampling spectrum. In this way the requirements placed on the steepness of the stop-band slopes of the pre- and post-filter are reduced accordingly, as already mentioned in 2.1.

A further advantage is the now possible partial relocation of the required signal filtering to the area of digital signal processing, where even structures of a higher order can be realised with a favourable effect – for instance, without phase distortion – and at low cost.

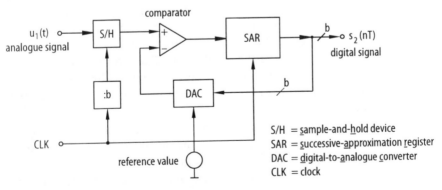

Fig. 2.6. ADC in accordance with the weighting process

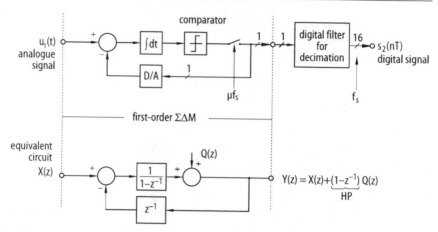

Fig. 2.7. ADC with a sigma-delta modulator

Finally, the oversampling can be utilised for the exchange of the speed and the resolution of a converter. As a topical application of the above-mentioned advantages of advanced semiconductor technology the sigma-delta modulator ($\Sigma\Delta M$), which has received wide acclaim in audio technology as an ADC, will be presented below.

The $\Sigma\Delta M$, the block diagram of which is shown in the upper part of figure 2.7, can be derived from the well-known delta modulator (ΔM) by changing the position of a module – the integrator – within the circuit [PARK, CANDY]. A bipolar signal, which has been derived from the binary output signal of the ADC, is subtracted from the actual analogue signal value $u_1(t)$. The resulting difference signal is fed via the integrator to a comparator, which acts as an ADC with a one-bit resolution. The actual sampling process can be accomplished through the activation of the comparator by the clock signal, which is represented here by a switch.

The whole circuit can be conceived as a sampled control loop keeping the continuously integrated difference between input and output signal as small as possible. As the output signal can only assume two states, it undergoes a pulse-density modulation (PDM) and thus a change of its mean value, which tries to follow the amplitude of the input signal. The error occurring on account of the quantisation with 1 bit can only be reduced by oversampling, which leads to a diminishing of the changes in the input signal between two successive clock periods.

As already mentioned, $\Sigma\Delta M$ and ΔM are similar in their structure and also in their function. However, the transfer of the integrator from the feedback loop (ΔM) to the forward loop ($\Sigma\Delta M$) leads to a different response in the frequency domain. In the equivalent circuit (figure 2.7, lower part) the $\Sigma\Delta M$ circuit is shown after undergoing a z-transformation with the integrator repre-

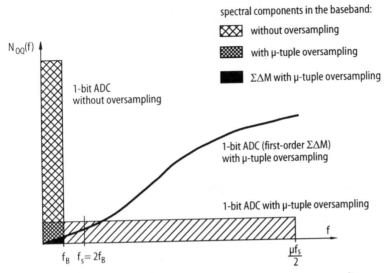

Fig. 2.8. Spectra of quantisation errors of 1-bit ADCs for audio signals with oversampling

sented by the operation $1/(1-z^{-1})$ [PARK, AGRAW]. At this point the integrator causes the quantisation error $Q(z)$ in the output signal $Y(z)$ to be weighted with a high-pass characteristic $(1-z^{-1})$, whereas the input signal $X(z)$ remains unweighted. As opposed to this, both components in the ΔM are unweighted in their frequency responses after demodulation.

The spectra of the quantisation errors in figure 2.8 show these effects. A 1-bit ADC in the baseband f_B with a Nyquist sampling frequency of $f_s = 2f_B$ distributes the power density N_{oQ} of its quantisation error evenly in this area with a correspondingly high value. Oversampling, for the same converter, results in a distribution, equal in area, over a μ-tuple-extended spectrum but with a correspondingly low value. After the filtering-out of the baseband the residual noise-power density is reduced by a μ-tuple, i.e. the oversampling factor. In the same way the ΔM functions as an ADC.

With the $\Sigma\Delta M$ as a 1-bit ADC it is possible to further reduce this noise-power density, as shown schematically in figure 2.8. This is due to the spectral noise shaping resulting from the high-pass characteristic [PARK], so that after filtering out the baseband, only a triangular-shaped area remains as noise-power density. However, this spectral noise shaping of the $\Sigma\Delta M$ of the first order cannot suffice for practical applications with an effective resolution of, for example, 16 bits. By cascading several such circuits it is possible to obtain higher $\Sigma\Delta M$ orders, which make the required noise reduction in the baseband possible [CANDY, GRAY].

As shown in figure 2.7, a digital filter follows the $\Sigma\Delta M$, which by decimation makes the required number of bits available in the output signal while simul-

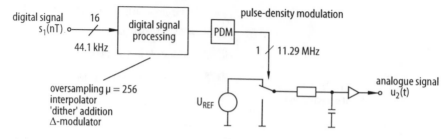

Fig. 2.9. 1-bit DAC for audio signals with oversampling

taneously reducing the external clock frequency to f_s. Usually this operation is carried out by a number of series-connected subfilters.

The DACs for audio signals often work on the same principle as those for video signals as depicted in figure 2.3. But the high-resolution requirement causes problems for the control of the resistor tolerances and the very small individual currents in the range of 10^{-8} A. Therefore different variants of this type of converter are in use which, due to internal splitting into several signal branches and by temporal averaging out of the errors which occur as a result ('dynamic element matching'), achieve a technically better solution [SKRI-TEK].

The use of an analogue signal as reference current I_{REF} is an interesting application of this DAC. In this case the converter functions as an attenuator, the attenuation of which is determined by the digital input signal. This arrangement is well-suited to a remote-controlled level- and volume adjustment. The 'fineness' of the step-by-step characteristic of such adjusting devices is determined by the number of bits b in the DAC.

As already mentioned, oversampling is also advantageous for DACs. The corresponding inverse application of the principle governing the ADC leads to a 1-bit DAC in accordance with figure 2.9 [BIAESCH]. To start with, a large number of new values between the present values of the input signal $s_1(nT)$ are generated by an interpolation process. By this method the differences between these interpolated values become smaller, so that a quantisation with a resolution of only 1 bit is sufficient in the case of a correspondingly large oversampling (here $\mu = 256$). As with the ΔM, a PDM signal is generated from the 1-bit sequence, which controls a switch as a 1-bit DAC. An additional 'dither' bit results in a perceivable reduction of the quantisation noise at very low signal levels. The subsequent analogue integrator smooths the form of the output signal $u_2(t)$. An advantage of this concept – apart from the easy implementation as an IC – is the relatively high linearity at small amplitudes.

Symbols in Chapter 2

B	colour component signal: blue
b	number of bits with which a signal is digitised
b_i	bit value i
C_B	digital colour difference signal: blue
C_R	digital colour difference signal: red
E'_{CB}	analogue colour difference signal: blue; γ pre-corrected
E'_{CR}	analogue colour difference signal: red; γ pre-corrected
E'_Y	analogue luminance signal, γ pre-corrected
f	frequency in general
f_B	limiting frequency of the baseband signal
f_s	sampling frequency
G	colour component signal: green
H_o	data rate of a single signal
H_{oTotal}	data rate of a total signal
H_{oSer}	data rate of a serial signal
I_{REF}	reference current
$i(t)$	analogue signal in the time domain (current)
n	running variable, integer
N_{oQ}	power density of quantisation-error signal
N_Q	r.m.s. value of quantisation-error signal
Q	size of one quantisation step
$Q(z)$	quantisation-error signal, transferred to the z plane
R	colour component signal: red
S_A	audio-signal amplitude (r.m.s. value)
S_V	video-signal amplitude (peak-to-peak value)
$s(nT)$	digital signal at sampling instants nT
T	clock period, inverse of f_s
t	time in general
U_{REF}	reference voltage
$u(t)$	analogue signal in the time domain (voltage)
μ	oversampling factor
$X(z)$	input signal, transferred to the z plane
Y	digital luminance signal
$Y(z)$	output signal, transferred to the z plane
z	variable of z-transformation

3 MPEG Source Coding of Audio Signals

The digital television of the future will also have digital associated sound. In this way an audio quality can be achieved which is far better than that obtained with the FM transmission system used in analogue television. However, in order to keep a digital television transmission within the realms of a realistic bandwidth, a suitable bit-rate reduction of the audio source signal is required. Moreover, the transition to a digital associated sound should offer the content providers the opportunity to choose between various data rates and thus between various qualities of audio signals. Furthermore, several alternative methods (such as dual-channel audio, stereo, surround-sound etc.) should be available to the content provider. It goes without saying that certain compatibility requirements must be satisfied so that each receiver is in a position to decode the audio signal regardless of which system or bit rate has been chosen by the content provider.

The techniques for the source coding of audio data which meet these requirements have been worked out by the MPEG audio group, a subsidiary group of the ISO/IEC JTC1/SC29/WG11. The standards developed by this group are referred to by experts as the MPEG standards and designated as IS 11172-3 (MPEG Audio) and IS 13818-3 (MPEG-2 Audio). According to the decisions taken by the DVB Project the coding of the associated audio in digital television must comply with the above standards to be discussed in this chapter. After a survey of the bit-rate reduction the psychoacoustic basics of the coding techniques will be explained. Section 3.3 will deal mainly with the coding and decoding in accordance with MPEG. Following that, there will be an overview of the coding of surround-sound in accordance with MPEG-2.

3.1 Basics of Bit-rate Reduction

Figure 3.1 shows the block diagram of the transmission of information with bit-rate reduction. The information source supplies an analogue audio signal which is transformed into a digital signal by sampling and quantisation. The usual studio sampling rate of 48 kHz and a 16-bit quantisation result in a bit rate of 768 kbits/s for a monophonic channel. In order to transmit this signal with the least possible energy input, the bit rate is further reduced in a source

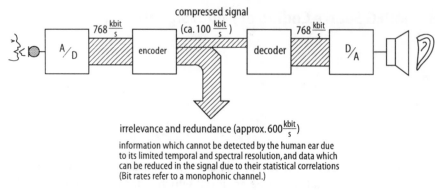

Fig. 3.1. Principle of bit-rate reduction for audio signals

coder. The decoder at the receiving end converts the bit-rate-reduced signal back into a PCM signal, which is then changed into an analogue signal, amplified, and used to drive a loudspeaker.

The art of source coding requires that the bit-rate reduction be carried out in such a way that the information sink (in the case of audio coding: the human ear) cannot detect any deterioration in the sound quality or that the negative effects caused by the bit-rate reduction are at least kept to a minimum. In the literature of information technology, two basic approaches are identified: redundancy reduction and irrelevance reduction.

The redundancy of a signal is a measure of the predictability of portions of that signal. To reduce the redundancy, one needs to know the statistical properties of the information source. Redundant signal portions can be reduced at the encoder and restored at the receiver so that the original audio signal can be completely reconstituted. However, generic audio signals – unlike specific ones such as speech signals – hardly ever contain the quantity of statistical correlation necessary for obtaining high compression rates by means of a simple redundancy reduction.

An irrelevance reduction, on the other hand, takes advantage of the limited capacity of the information sink in order to compress the data. In the case of audio signals those signal portions which the human ear – owing to its limited capacity for discrimination – cannot perceive in amplitude, time and spectrum, are eliminated in the encoder. Contrary to the redundancy reduction, the irrelevance reduction is an irreversible process. However, since this reduction cannot be perceived by the human ear, the ear is not aware of the degeneration of the signal. As an irrelevance reduction takes advantage of the limited receptivity of the sink, this type of compression of audio data requires a thorough understanding of the perception limits of the human ear. For this reason some findings with regard to the capacity of the human ear will be discussed in the following.

3.2 Psychoacoustic Basics

First of all, the perception of stationary sounds will be described in which no influence of the time structure is ascertainable. This is the case for sound with a duration of not less than 200 ms [ZWICKER]. In this section the threshold of audibility and the auditory sensation area, which describe the perception of single sinusoidal tones, will be explained. If a sound event is comprised of more than one sinusoidal tone, individual sounds may be masked by others. This masking effect is explained in section 3.2.2. Finally there is a description of time-dependent events in the perception of sounds.

3.2.1 Threshold of Audibility and Auditory Sensation Area

The threshold of audibility is that sound-pressure level of a sinusoidal tone that can just be perceived. It is dependent on the frequency. The sound pressure level L can be defined as follows:

$$L = 20 \log\left(\frac{p}{p_0}\right), \tag{3.1}$$

where p represents the sound pressure and p_0 can be defined as:

$$p_0 = 2 \cdot 10^{-5}\, \frac{N}{m^2} = 20\,\mu Pa. \tag{3.2}$$

In order to determine the threshold, test subjects were required – in a subjective study – to set the sound level of a sinusoidal source to the point at which it just ceased to be perceptible. This sound-pressure level was then recorded, determined for various frequencies and averaged over a large number of subjects. These levels form the threshold of audibility shown in figure 3.2[1]. The threshold of pain was reached at a sound-pressure level of 130 dB. The range between the threshold of audibility and the threshold of pain is called the auditory sensation area. Further, curves of the same loudness perception can be seen in figure 3.2. These curves show that the loudness perception and the threshold of audibility depend to a great extent on the frequency, which is an indication that for audio coding it is recommendable to treat different parts of the spectrum differently.

However, most sounds which occur in nature are not comprised of isolated sinusoidal tones. The perception by the human ear of more complex sounds must therefore be examined.

---

[1] To be precise, the median of all measured values is taken. The reason for this is explained in [ZWICKER].

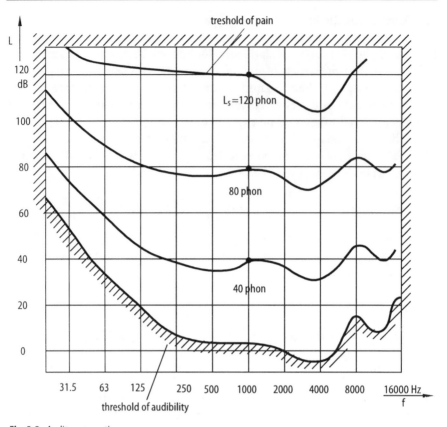

Fig. 3.2. Auditory sensation area

3.2.2 Masking

Masking is an effect known from everyday life. For instance, an audio event (such as music) which is easily heard in a quiet environment can be imperceptible in the presence of noise (as that caused by a pneumatic drill). In order for the useful sound to remain perceptible in spite of the noise it is necessary for the level to be considerably higher than in a quiet environment. The concept of "masking thresholds" is used for the quantitative description of the masking effects. A masking threshold is defined as the sound-pressure level to which a test sound (as a rule, a sinusoidal test tone) must be reduced to be only just audible beside the masking signal [ZWICKER]. The masking thresholds and the method of their determination are described below.

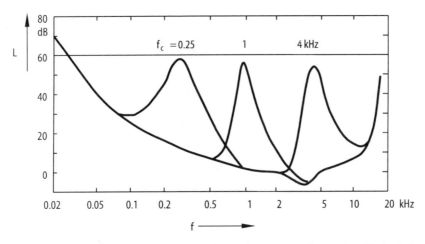

Fig. 3.3. Masking threshold caused by narrow-band noise of varying centre frequencies with a level of $L_G = 60$ dB according to [ZWICKER]

3.2.2.1 Masking by Stationary Sounds

First of all, we will discuss the masking by sound whose time structure can be disregarded. The time structure of a sound can be disregarded if the duration of the sound is longer than 200 ms. In order to define the masking threshold caused by narrow-band noise, the subjects are exposed to a sinusoidal tone in addition to the narrow-band noise. The sound-pressure level of the sinusoidal tone is progressively reduced until it becomes just inaudible. This procedure is repeated with several hundred test subjects for a range of frequencies between 20 Hz and 20,000 Hz. The determined sound-pressure levels – averaged over a number of subjects – constitute the masking threshold.

Figure 3.3 depicts the masking threshold by means of a masking narrow-band noise level of 60 dB, at centre frequencies of 250 Hz, 1 kHz and 4 kHz with bandwidths of 100 Hz, 160 Hz and 700 Hz resp. All tones at a level below these thresholds are masked by the narrow-band noise and are therefore imperceptible to the human ear. As can be seen, the course of these masking thresholds is highly dependent on frequency. This entails that the course of the masking thresholds depends on the centre frequency of the narrow-band noise and that the distance of the maxima from the 60-dB line increases with increasing frequency. However, the frequency of the masker is not the only influence on the form of the masking threshold. Another parameter is the level of the masker. The masking threshold, caused by narrow-band noise of varying sound-pressure levels L_G at a centre frequency of 1 kHz and a bandwidth of 160 Hz, is shown in figure 3.4. Apart from the frequency and the level of the

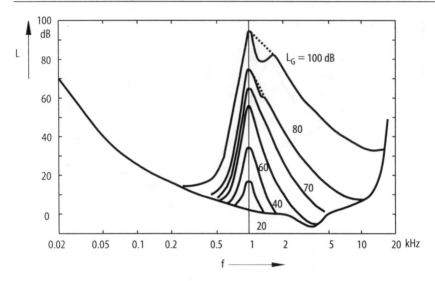

Fig. 3.4. Masking threshold caused by narrow-band noise of varying levels according to [ZWICKER]

masker, a third parameter has to be taken into account, namely the tonality of the masker. If the masking tone is sinusoidal, the resulting masking threshold will differ from that depicted in figure 3.4.

Until now, only the masking by stationary sound has been assessed, i.e. by sinusoidal tones or narrow-band noises with a long duration. In the following section the effects observed in sound events of a very short duration will be discussed.

3.2.2.2 Time-dependent Masking Effects

In this section we will examine the question to which degree masking effects occur in maskers of a very short signal duration. In order to determine this, subjects were each exposed to a masking impulse of white Gaussian noise with an amplitude of L_{WN} for a duration of 0,5 s. This masking impulse was followed, at an interval of t_v, by a pressure impulse of a duration of 20 µs (see figure 3.5a). The subjects had the task of adjusting the amplitude of this Gaussian impulse so that it was just imperceptible. The results of this subjective study are shown in figure 3.5b. The masking threshold still has approximately the same level, up to about 10 ms after switching off the masking noise, as when the masking noise and the Gaussian impulse are presented simultaneously. It is only after these approx. 10 ms that the masking threshold falls, reaching the threshold of audibility after approx. 200 ms.

However, masking not only occurs when the Gaussian impulse is offered during or after the masker but also prior to it. This effect – in the literature de-

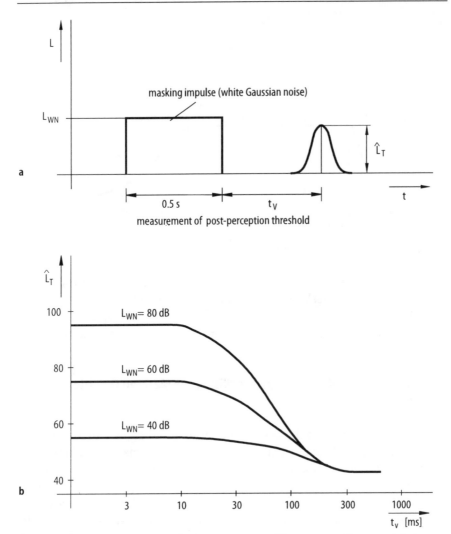

Fig. 3.5. Post-masking according to [ZWICKER]. **a** Masker and test impluse; **b** Post-hearing threshold $\hat{L}_T$ at different amplitudes of the masker and different distances to the masker

scribed as pre-masking – is only detectable when the length of the interval between the Gaussian impulse and the switching on of the masker is no more than 20 ms.

The masking effects are summarised in figure 3.6. The masking of sound events which are presented to the human ear at the same time as a masker are also referred to as 'simultaneous masking'. However, the masking effect remains after the masker has been switched off and disappears after approx. 200 ms have elapsed. This is referred to as 'post-masking'. Even sounds pre-

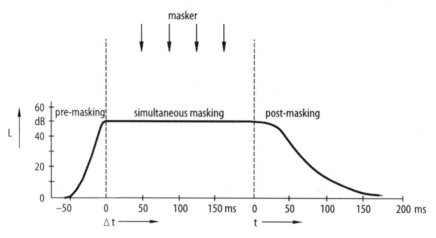

Fig. 3.6. Summarisation of masking effects

sented before the masking noise commences can be masked. But this effect, which is called 'pre-masking', subsides after only about 5 ms.

In the following sections we shall show that these masking effects can be exploited for a very effective irrelevance reduction in audio signals.

3.3 Source Coding of Audio Signals Utilising the Masking Qualities of the Human Ear

This section will discuss the source coding of audio signals in accordance with MPEG standards [ISO 11172] and [ISO 13818]. The explanation of the basic structure of an MPEG audio coder will be followed by the description of the bit-rate reduction in accordance with MPEG layer 1. Layer 1 was developed with the aim of keeping the implementation requirement for the encoder to a minimum. There will then follow a description of layer 2 which compared to layer 1 is more complex but enables higher compression of an audio signal while retaining the same quality. To complete the picture, the third layer in the MPEG standard will be briefly dealt with. The compression rate in layer 3 is higher again than that of layer 2, involving a yet higher implementation requirement. However, the "Guidelines on the implementation and usage of Service Information (SI)" of the DVB Project [ETR 211] make no provision for the use of layer 3 in digital television.

Those developing the MPEG standards took care to greatly reduce the expenditure involved in the decoding of audio signals, as compared to the cost incurred in the encoding process. In distribution services, such as TV broadcasting, low expenditure is of great importance so as to make possible an economical manufacture of the receiving equipment. The decoding of the bit

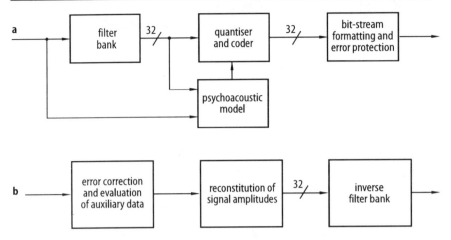

Fig. 3.7. Principle of bit-rate reduction in accordance with the MPEG standard. **a** Encoder; **b** Decoder

stream is described in section 3.3.5. To conclude, the audio coding in accordance with MPEG-2 is outlined. This is a coding for surround-sound, which is both forward and backward compatible with the standard for MPEG audio.

In order to ensure compatibility between receivers from different manufacturers two items only need to be laid down in the standard, viz, the decoding procedure and the shape of the resulting signals. In contrast to this, the description of the MPEG coder in the standard is of a purely informative nature and the design may vary from one manufacturer to another.

3.3.1 Basic Structure of the MPEG Coding Technique

The basic structure of an MPEG audio coder is shown in figure 3.7a. The audio signal to be coded is first passed through a filter bank, which breaks it down into 32 frequency bands of equal bandwidth. The ISO/MPEG standard provides for the use of a polyphase filter bank, the coefficients of which are listed in the standard. Simultaneously in each of these 32 channels the signal is critically subsampled, which means that the sampling rate is reduced to the thirty-second part of the sampling rate used in the digitisation. These signals are then fed into a quantiser which quantises the signals in each individual band in such a way that, whilst the number of steps is kept as low as possible, the quantisation noise still remains below the masking threshold so that noise is not perceptible to the human ear. It goes without saying that the permissible degree of quantisation cannot be determined without a proper knowledge of the psychoacoustic laws described in section 3.2. A particular element of the coder – called 'psychoacoustic model' in figure 3.7a – evaluates the input signal and, taking into account the masking effects, computes the permissible quantisation for each subband. Subsequently, the quantised sam-

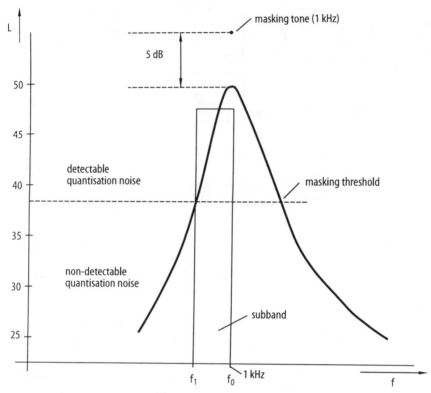

Fig. 3.8. Derivation of the optimal quantisation threshold according to [THEILE]

ples and a number of auxiliary data which are required for the reconstitution of the audio signal by the decoder are formatted into a bit stream and furnished with an error protection.[2]

The operations at the decoder (figure 3.7b) start with a correction of transmission errors, followed by an interpretation of the auxiliary data and a reconstitution of the samples of the audio signals. The reconstituted 32 subbands are then conveyed to an inverse filter bank and there combined into one frequency band.

The achievable compression ratios depend to a great extent upon a suitable determination of the quantisation thresholds in each subband. Therefore the determination of these thresholds shall here be described in detail. First, a 1-kHz tone is assumed as masker (figure 3.8). The most critical case exists when this tone is to be found in the upper band limit of the corresponding subband, since the masking thresholds show a steeper curve when coming from low frequencies. In this case the masking effects in the subband in ques-

---

[2] The error protection is optional.

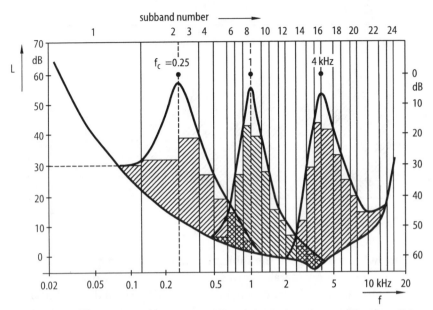

Fig.3.9. Amplitude spectrum with masking thresholds and quantisation noise according to [STOLL 1]

tion are lowest. The intersection of the masking threshold with the lowest band limit now defines a level underneath which no audio signals are perceptible, so that these need not be reproduced at the decoder. Therefore the quantisation of this subband can be chosen roughly enough up to the point at which the quantisation noise reaches this value (broken line in figure 3.8).

However, in general we are not concerned with one single masker, but with a music or speech signal which is far more complex. Figure 3.9[3] shows an example of three sinusoidal tones which function as maskers. These do not only mask signal parts in the subbands in which the masking noise itself is detected, but also in neighbouring bands. In figure 3.9 the auditory sensation area which is masked by these three tones is hatched in. The quantisation noise may cover this area without it being perceptible to the listener. In individual subbands, under particular circumstances, even all signal parts may lie below the masking threshold, so that these do not need to be transmitted. It is apparent from figure 3.9 that the aim of coding is to shape the spectrum of the quantisation noise so that it is masked by the signal. With the aid of this spectrum-shaping – in contrast to white Gaussian noise – far more quantisation noise is permissible without its becoming perceptible to the human ear.

---

[3] A representation of 32 bands of the same width, as specified in the MPEG standard, leads to an unclear picture due to the logarithmic abscissa. Therefore the diagram does not show 32 bands of equal width.

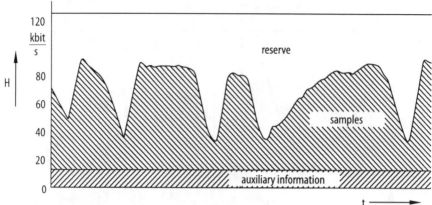

Fig. 3.10. Dynamic bit-flow reserve

Therefore, with no change in the subjective quality of the signal, this type of encoder requires a considerably lower bit rate than a PCM encoder.

From figure 3.9 it can also be seen that the extent to which the quantisation noise is masked by the audio signal, and therefore imperceptible, depends heavily on the amplitude and spectral distribution of the masking noise itself. However, the permissible quantisation noise also indirectly determines the required bit rate of the signal so that the bit rate is bound to vary as a function of the extent of the masking. Figure 3.10 shows the required time-dependent bit rate H in the case of a natural microphone signal. On the one hand, the bit stream contains the auxiliary information necessary for the reconstitution of the subbands and for controlling the decoder. The bit rate required for this is almost constant. On the other hand, the bit stream contains the quantised samples, the bit rates of which are very much dependent on the quantisation noise allowed. As most information channels are defined by a constant maximum permissible bit rate, one would give away part of the available capacity when using a signal with a time-dependent data rate. For this reason the quantisation steps are always increased to a number which ensures that the capacity of the channel is completely utilised. Therefore the maximum values for the quantisation noise, as shown in Figure 3.9, can always be underrun. The advantages of this approach are explained below.

If a quantisation-noise signal is allowed to reach the masking threshold, as in figure 3.9, then no further processing of this signal can take place without quantisation noise becoming perceptible. This is even true for a simple change in the frequency response by means of an equaliser. If the frequencies of the spectrum shown in figure 3.9 were to be pre-emphasised upwards of 5 kHz, the quantisation noise would exceed the masking threshold. The same considerations apply when a signal which has been coded in this form is decoded and then recoded ("cascading"). This recoding is accompanied by a

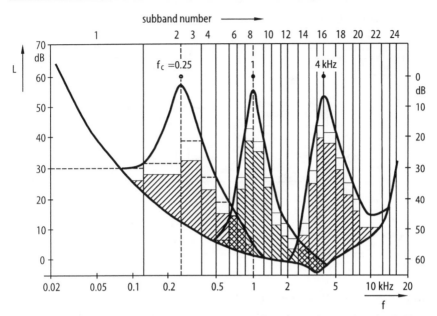

Fig. 3.11. Utilisation of the bit-flow reserve for maximisation of the distance from masking threshold to quantisation noise in accordance with [STOLL 1]

new quantisation and therefore by an increase in the quantisation noise above the masking threshold.

In order to allow a further processing of the signal – at least to a certain extent – the quantisation is fixed in a way that there is a margin between the masking threshold and the quantisation noise (figure 3.11[4]). For this, the margin is increased in an iterative process until the maximum available data rate of the channel has been exhausted. Therefore, the larger the bit rate the greater the post-processing capacity of the signal.

Following the discussion of the basic principles of psychoacoustic coding according to the MPEG standard, the MPEG encoder and decoder will now be described in detail.

3.3.2 Coding in Accordance with Layer 1

Figure 3.12 shows the block diagram of an audio coder in accordance with MPEG layer 1. The filter bank which divides the signal into 32 subbands of equal bandwidth has already been described in section 3.3.1. In these subbands a critical subsampling of the signal takes place. Subsequently, the maximum value – the scaling factor – of a block of 12 samples is determined, the

---

[4] See footnote 3.

digital audio signal (PCM)
(768 kbit/s)

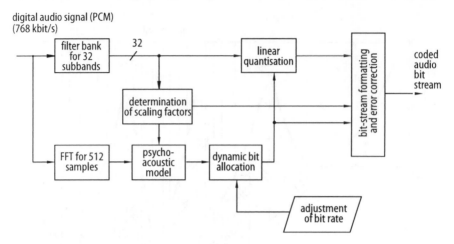

Fig. 3.12. Block diagram of a coder in accordance with layer 1 of the MPEG standard (mono)

samples corresponding to 8 ms signal duration at a 48-kHz sampling rate. This scaling factor is quantised and transmitted with an amplitude resolution corresponding to 6 bits. The 12 samples are then divided by the scaling factor, and the resulting signal is subjected to quantisation. For computing the required number of quantisation steps one needs to apply the masking threshold provided by the psychoacoustic model. All 12 samples are then subjected to the same quantisation. This is permissible because, due to the pre- and the post-masking, the effect of a masker extends over the entire period in question, which lasts between 8 and 12 ms, depending on the sampling rate.

In parallel with the subband analysis the signal is subjected to a Fourier transform with 512 samples, so that an even higher resolution is obtained for the spectrum of the signal. The local maxima are then determined in the spectrum, and the neighbourhood of these maxima is evaluated in order to determine whether these are tonal or non-tonal components of the signal. This evaluation is necessary since the tonality of a signal has a considerable impact on the shape of the masking thresholds (see section 3.2.2.1). In this way it is possible to determine, from the masking thresholds, the maximum sound level in the subbands.

On the basis of the masking thresholds and the given bit rate the number of the quantisation steps can be so determined as to maximise the margin between the masking threshold and the quantisation noise. The quantised samples are combined with the scaling factors and the bit allocation to form a bit stream ("bit-stream formatting"), the bit allocation indicating the number of bits for each sample.

These data are preceded by a header which contains information for the controlling of the decoder. Figure 3.13 shows the construction of the bit

data frame with 384 PCM samples
(equal to 8 ms at a 48-kHz sampling frequency)

header	error protection	bit allocation	scaling factors	samples	auxiliary data

| 12 bits synchronising signal and 20 bits system information | 16 bits (optional) | 4 bits each | 6 bits each | 2 ... 15 bits each | |

Fig. 3.13. Audio bit stream in accordance with layer 1 of the MPEG standard (mono)

stream at the output of an MPEG-layer-1 encoder. The data frame shown is defined as part of the stream which contains all the relevant information required for the decoding. At the front of the frame is the 32-bit header, which starts with a synchronising word and contains the subsequent system information describing the audio signal in detail. A following error protection is optional and serves to protect the most important data, the bit allocation, and part of the header against transmission errors. A convolutional code [ISO 11172] has been provided for this error protection. Next, there follows the bit allocation, with the number of bits required for each of the samples, and the scaling factors, which are quantised with 6 bits each. The samples as such are represented by between 2 to 15 bits according to the permissible quantisation noise. There can be further data added to the frame which are not necessarily interpreted by an MPEG decoder. Among other things, a compatible extension leading to a surround-sound system is made possible by these auxiliary data. This possibility will be dealt with in section 3.3.7.

3.3.3 Coding in Accordance with Layer 2

Contrary to the layer-1 encoder, the layer-2 encoder (figure 3.14) combines 36 PCM samples into a block, which, at a sampling frequency of 48 kHz, corresponds to a section of 24 ms of the audio signal. In this case one scaling factor per subband might prove to be no longer sufficient, as the effect of the premasking only lasts for a maximum of 20 ms. This must be taken into account in the case of major temporal changes in the signal ("drum beats"), which would require two or three scaling factors per subband and per block. But if the signal is not subject to major temporal changes, one scaling factor per subband will suffice. There is a selection unit in the encoder which determines the required number of scaling factors. The next step consists in determining the maxima of the samples of a block. These values form the scaling factors. Apart from the scaling factors themselves, information about their number has to be transmitted the decoder.

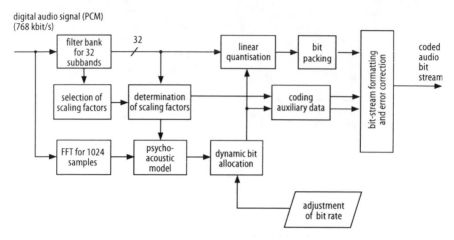

digital audio signal (PCM)
(768 kbit/s)

Fig. 3.14. Block diagram of a coder in accordance with layer 2 of the MPEG standard (mono)

To determine the tonality of the maskers, a Fourier transform on 1024 samples is carried out in the case of layer 2 so that a higher spectral resolution is obtained than with layer 1. The determination of the masking threshold and the formatting of the bit stream are both performed in the same way as in layer 1.

The fixing of the number of quantisation steps also differs from that in layer 1. In layer 2 it is only in the lower subbands that all values between 2 and 15 bits are admissible for the representation of the samples. In the upper subbands the possibilities of representation are limited. Thus, at a bit rate above 50 kbit/s and at a sampling rate of 48 kHz, only 0, 3, 5 or 65,535 quantisation steps are admissible in subbands 23 to 26. Therefore, as only four different quantisations are possible, only 2 bits are required for the transmission of the bit allocation.

As the signal energy in the upper bands is generally very low, a greater number of quantisation steps is seldom required. The required data rate is therefore kept to a minimum in that only 0, 3, 5 or 65,535 quantisation steps are allowed and only 2 bits are thus required for the bit allocation. At a sampling rate of 48 Hz the subbands 27 to 31 are not transmitted since the frequencies involved are above 20 kHz and are therefore no longer audible.

Figure 3.15 shows the frame of an MPEG-layer-2 bit stream. This differs from that of layer 1 only in that it comprises 1152 bits and that in addition it contains an identification of the number of scaling factors used in the bit stream.

3.3.4 Coding in Accordance with Layer 3

To complete the picture, the coding in accordance with MPEG layer 3 should also be presented. This is an algorithm developed with the aim of increasing

frame with 1152 PCM samples
(equal to 24 ms at a 48 kHz sampling frequency)

header	error protection	bit allocation	label for number of scaling factors	scaling factors	samples	auxiliary data
12 bits synchronising signal and 20 bits system information	16 bits (optional)	4 bits each for lower 3 bits each for middle 2 bits each for upper subband	2 bits each	6 bits each	2 ... 15 bits each	

Fig. 3.15. Audio bit stream in accordance with layer 2 of the MPEG standard (mono)

the compression as compared to layer 2, which, however, entails a higher implementation requirement. Figure 3.16 shows the block diagram of the coder.

Layer 3 requires an additional division of the signal spectrum into 576 sub-bands, which is achieved by means of a modified discrete cosine transformation (MDCT). Since the pulse duration of a filter increases the narrower the chosen bandwidth, this additional division of the frequency band has the disadvantage that the resolution of the coder in the time domain is reduced to an unacceptable level. Therefore a signal-adaptive changeover from high time-resolution to low frequency-resolution and vice versa is provided. This ensures a more precise definition of the masking thresholds and therefore a better adaptation of the quantisation noise to the limits of perception.

Further, layer 3 provides for a non-linear quantisation, which results in a favourable adaptation to the human ear, since the latter also shows a non-

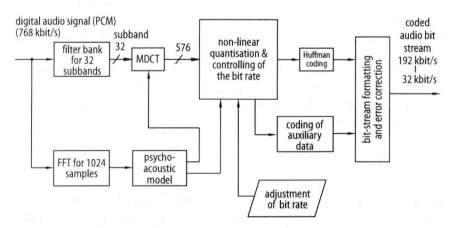

Fig. 3.16. Block diagram of a coder in accordance with layer 3 of the MPEG standard (mono)

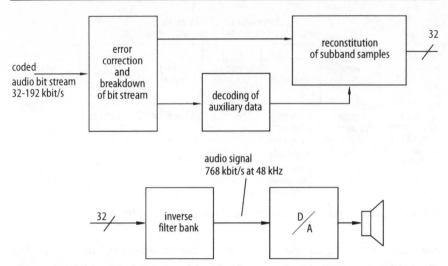

Fig. 3.17. Block diagram of a decoder in accordance with layers 1 and 2 of the MPEG standard (mono)

linear behaviour with regard to the perception of changes in sound level [ZWICKER].

By using a Huffman encoding procedure the redundancy still contained in the signal can be reduced so that an even lower bit rate can be achieved while maintaining an audio signal of good quality.

3.3.5 Decoding

At the decoder, first the error correction is performed and then the bit stream is broken down into the samples and the auxiliary information (figure 3.17). Through the evaluation of the auxiliary information (bit allocation, scaling factors) the individual samples can be reconstituted for the 32 subbands. By means of an inverse filter bank the subbands can then be combined again.

Since there is no psychoacoustic model to be evaluated at the decoder, the technique is much simpler than at the encoder. As noted above, it is of great importance, most particularly for radio distribution services, to have a simple decoding technique.

3.3.6 The Parameters of MPEG Audio

In this section the most important parameters for MPEG audio coding will be summarised. In accordance with the implementation guidelines of the DVB Project [ETR 211] a receiver must have an MPEG-compatible audio decoder which supports all operational modes stipulated in the standard. An MPEG decoder must be able to decode mono signals and stereo signals as well as two completely independent channels ("dual-channel sound"). A further operational mode is joint stereo. This mode is based on the knowledge that in intensity stereophony for frequencies above 2 kHz only the envelope and not the fine structure of the signal contributes to the stereo effect. For this reason one can concentrate on the coding of the aggregate signal for frequencies above 2kHz and need only transmit the scaling factors for both channels.

The sampling rates recommended for the audio signal to be transmitted are 32 kHz, 44.1 kHz and 48 kHz. The total bit rate of a coded audio signal lies between 32 kbit/s and 384 kbit/s for layer 1, and between 32 kbit/s and 448 kbit/s for layer 2. In the standard there are 14 bit rates each defined within these limits which must be supported by every decoder.

Further, a decoder conceived for layer 2 must also be able to decode a layer-1 bit stream. Accordingly, a decoder for layer 3 must be compatible with layers 1 and 2.

The operational modes and data rates described must be supported by an MPEG-compatible decoder and therefore also by a decoder for a digital television set ("integrated receiver decoder"). The content provider thus has considerable freedom to determine the audio coding to be used and therefore also the type and quality of the audio signal.

Apart from the above-mentioned operational modes a content provider can also offer surround-sound. In November 1994, during the second phase of the MPEG activities, a separate international standard was adopted for the coding of a surround-sound signal. This will be outlined in the following section.

3.3.7 MPEG-2 Audio Coding

A loudspeaker configuration with one central loudspeaker and two surround loudspeakers – in addition to the usual left and right channels for stereo – is specified by MPEG for the surround-sound system. As three frontal and two surround loudspeakers are used, this arrangement is referred to as 3/2 stereo. Figure 3.18 shows the loudspeaker configuration recommended by ITU-R in [CCIR 10].

During the development of the coding system great emphasis was placed on backward compatibility, which means that a surround signal must be decodable by an MPEG decoder – of course as a stereo signal. This backward compatibility makes it possible for a content provider to transmit surround-sound, and for viewers with simple MPEG decoders and no surround-sound decoders to have the programme decoded (though only in stereo quality). Moreover, the MPEG-2 audio coding technique is forward compatible, which means that an MPEG-2 decoder can also decode an MPEG data stream.

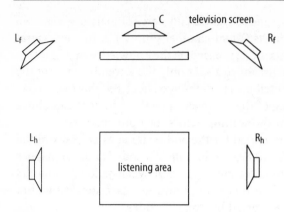

Fig. 3.18. Loudspeaker configuration for a 3/2 stereo system

To ensure backward compatibility both stereo signals L_0 and R_0 are computed from the 5 channels L_f, R_f, L_h, R_h, and C by a linear equation as follows:

$$L_0 = L_f + xC + yL_h$$
$$R_0 = R_f + xC + yR_h \; .$$

The coefficients x and y typically have the value 0,71 [STOLL 2].

Compatibility can now be achieved by formatting the bit stream of a surround signal in accordance with that of layer 2 (figure 3.15), where the channels L_0 and R_0, which constitute the compatible stereo signal, are placed at the front part of the frame, whereas the other three channels are transmitted in that part of the frame which is used for auxiliary data. The MPEG decoder does not evaluate these auxiliary data and only decodes the stereo signal. The surround decoder, however, recognises from the header that the signal possesses further surround channels and evaluates these.

By means of a suitable data-rate reduction which also reduces redundancies and irrelevancies between the channels it is possible to code a surround signal without the data rate increasing to five times that required for a mono channel.

3.4 Summary

The sound for digital television is subjected to a bit-rate reduction in accordance with the MPEG standard. This coding technique reduces the amount of data by exploiting the limited spectral and temporal discrimination of the human ear. The technique is based on a breakdown of the spectrum into 32 subbands and on a block formation with 12 or 36 samples. For each of these blocks quantisation is carried out on the basis of a psychoacoustic model by exploiting the masking quality of the human ear. This quantisation can be performed such that the generated quantisation noise is shaped, in terms of spectrum and time, in such a way as to be masked by the signal and therefore to remain inaudible. At the output of the coder the bit stream is formatted so that the decoder can reconstitute the samples.

With MPEG-layer-2 audio coding it is possible to achieve a bit-rate compression to approx. 100 kbit/s without coding artefacts becoming perceptible (transparent quality). Furthermore, the various operational modes and data rates to which an MPEG audio decoder is suited ensure a significant adaptation of the encoding parameters to the requirements of the user.

MPEG layer 3 was developed with the aim of obtaining higher compression than with layer 2, which, however, increases the implementation requirement. Joint stereo coding permits an exploitation of redundancies between the two stereo channels. Audio coding in accordance with MPEG 2 constitutes a forward and backward compatible extension of the MPEG audio standard for coding surround-sound.

Symbols in Chapter 3

C	centre channel
f	frequency in general
f_c	centre frequency
f_l	lower cut-off frequency
f_u	upper cut-off frequency
H	data rate
L	noise-pressure level
L_f	front surround signal, left
L'_G	noise-pressure level for narrow-band noise
L_h	back surround signal, left
L_o	stereo signal after matrixing, left
L_s	loudness
L_T	noise-pressure level for a post-hearing threshold
L_{WN}	noise-pressure level for white Gaussian noise
p	noise-pressure level
p_0	reference noise pressure
R_o	stereo signal after matrixing, right
R_f	front surround signal, right
R_h	back surround signal, right
t	time in general
t_v	time interval
x	matrix coefficient
y	matrix coefficient
Δt	time interval

4 JPEG and MPEG Source Coding of Video Signals

In state-of-the-art television a digitised video signal typically uses a net bit rate of 166 Mbit/s (cf. section 2.2). If this data rate were transmitted without being compressed, considerably more bandwidth would be required than for the present analogue procedure. That is why the use of data compression techniques is indispensable for the development of a digital transmission standard for television signals. As with audio signals, a mere redundancy reduction (cf. section 3.1) would only result in a small average compression factor. With the aid of an irrelevance reduction which takes into account the characteristics of the human visual system, rejecting all imperceptible image contents, considerably higher reduction factors can be attained (with no change in the subjective quality of the image). This decreases the bandwidth requirement far below that for analogue transmission of a similar quality, thus making this procedure commercially very interesting.

A series of standards has been established for the coding of images, as shown in table 4.1 in historical order. First, the so-called JPEG (Joint Photographic Experts Group) standard was developed as ISO/IEC IS 10918 [ISO 10918], which was conceived for the efficient storage of still pictures [WALLACE, PENNEBK]. Owing to the early availability of reasonably priced ICs, a further utilisation offered itself for moving images. However, this "Motion JPEG" (M-JPEG) was not standardised and does not warrant compatibility of equipment purchased from different manufacturers.

Table 4.1. International standards for image coding

Standard	Range of application	Data rate
ISO/IEC IS 10918 "JPEG"	storage of stills; Motion JPEG: studio applications	not defined
ITU-T H.261 "p64"	ISDN, video conferencing	p * 64 kbits/s
ISO/IEC IS 11172 "MPEG-1"	CD-ROM, multimedia	up to 1.5 Mbit/s
ISO/IEC IS 13818 "MPEG-2"	television transmissions, studio applications	MP@ML: up to 15 Mbit/s 4:2:2@ML: up to 50 Mbit/s

By comparison, the standard ITU-T H.261 [ITU H.261], also known in some parts of America under the name of "p64", was conceived from the outset for the coding of moving images. It was optimised for visual telephony and similar applications, which can be offered via narrow-band ISDN networks at data rates of 64 kbit/s or integral multiples thereof.

With ISO/IEC IS 11172 [ISO 11172] MPEG-1 a technique was developed which was conceived for application in the field of multimedia [HUNG] where only a limited storage capacity or data rate is available (particularly for CD-ROMs) and therefore quality requirements have to be modest. Compromises that had to be reached in this matter include the limitation to the SIF format (half the spatial resolution compared to ITU-R BT.601 resolution), progressive sampling, and the 4:2:0 chrominance format (see section 2.2).

The extension of MPEG-1 in the direction of higher quality and, connected with this, higher data rates, is the MPEG-2 standard (ISO/IEC IS 13818 [ISO 13818]), which enables the transmission and storage of television signals based on interlaced scanning [KNOLL, TEICHNER].

The image-coding standards relevant to digital television will be discussed in the following sections.

4.1 Coding in Accordance with JPEG

4.1.1 Block Diagram of Encoder and Decoder

Figure 4.1 shows the block diagram of a JPEG encoder which is here discussed in detail. First of all, the input image is divided into blocks of 8×8 pixels. A two-dimensional discrete cosine transformation (DCT) which transforms the sampling values of the image into the so-called spatial frequency domain is applied to these blocks. Here the individual frequency components of the image appear in the form of coefficients, so that the subsequent quantisation can be carried out in varying degrees of coarseness, graded according to spatial frequencies, in order to enable an adaptation to the human visual system. The quantisation is the only lossy step in the algorithm (irrelevance reduction); all other steps (apart from rounding errors during computation of the

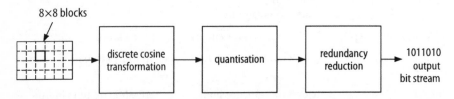

Fig. 4.1. Block diagram of the JPEG encoder

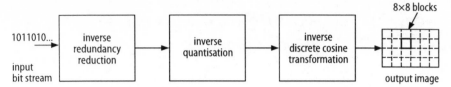

Fig. 4.2. Block diagram of the JPEG decoder

DCT) are completely lossless and can be revoked by the respective inverse operation. In order to further reduce the data rate, the quantisation is followed by a redundancy reduction which is essentially a combination of run-length coding and Huffman coding (see section 4.1.4).

The corresponding inverse processing steps, as shown in figure 4.2, are performed in the decoder. With the correctly chosen quantisation table (see section 4.1.3), the output image, which is constructed by putting the 8×8 pixel blocks back together, barely differs visually from the original image.

4.1.2 Discrete Cosine Transform

The discrete cosine transform can be conceived as a transformation of the original block of 8×8 pixels in the spatial domain into an equally sized block with 64 coefficients in the spatial frequency domain. Figure 4.3 shows such a transformation using an example taken from [HUNG]. The circular area, which can be conceived as a "pizza", with pixels differing in value from those surrounding it, can no longer be recognised as such in the spatial frequency domain (result of the DCT, in this case rounded to integers). It can be seen that the energy of the transformed block is concentrated at low frequencies. The higher the spatial frequencies f_x and f_y the smaller, in general, the coefficients. Apart from this, many coefficients are equal to zero, which proves very helpful for the data reduction.

Using the formula for a two-dimensional DCT with 8×8 pixels,

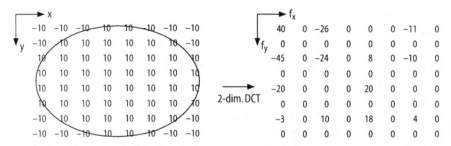

Fig. 4.3. Discrete cosine transformation (DCT) of an example 8×8 pixel block

$$G(f_x, f_y) = \tfrac{1}{4} C(f_x) C(f_y) \sum_{x=0}^{7} \sum_{y=0}^{7} g(x, y) \cos\left((2x+1) f_x \tfrac{\pi}{16}\right)$$

$$\times \cos\left((2y+1) f_y \tfrac{\pi}{16}\right)$$

(4.1)

where $C(f) = \begin{cases} \tfrac{1}{\sqrt{2}}, & \text{if } f = 0 \\ 1, & \text{if } f > 0 \end{cases}$

with: f_x, f_y = spatial frequencies
    $G(f_x, f_y)$ = DCT coefficients
    $C(f)$ = constant
    x, y = spatial co-ordinates
    $g(x, y)$ = video signal

we see that if the input signal $g(<x, y)$ is quantised with 8 bits, 11 bits would be required for the integral representation of the DC coefficient $G(0,0)$ (if it is rounded to an integer, which is equivalent to the finest possible quantisation with a quantisation step size = 1 in the subsequent step; see section 4.1.3). This is due to the adding up of the 64 sampling values $(\cos(0) = 1)$ and the scaling with the constant $\tfrac{1}{8} = \tfrac{1}{4} C(0) C(0)$. All other coefficients can equally be represented by 11-bit integers, since the values of the cosine functions in connection with the scaling factor or in front of the summation sign warrant the adherence to the 11-bit range of values. The transformation of 64 integers with an 8-bit precision to 64 integers with an 11-bit precision does not, of course, represent a decrease in the bit rate; on the contrary, it constitutes an increase. However, taking into consideration that the majority of the coefficients are very small and that the coarser quantisation in general leads to a narrowing of the range of numbers, a considerable decrease in the amount of data will be achieved in the course of the subsequent redundancy reduction.

Figure 4.4 illustrates the basis functions of the DCT. It shows blocks which, subsequent to the DCT, produce precisely one coefficient not equal to zero. The blocks are arranged according to the spatial frequency of this coefficient, which increases to the right and downwards. In this way the left uppermost block consists of only one uniform grey-scale value, which is generated by $G(0,0) \neq 0$. $G(f_x, f_y) = 0$ holds for all other f_x, f_y. Correspondingly, $G(7,7) \neq 0$ holds for the block in the bottom left corner. Each random block of a real image can be compiled by overlaying the 64 DCT basis functions with varying amplitudes. The coefficients $G(f_x, f_y)$ indicate just these amplitudes (including preceding sign).

The DCT is not the only transform which can be used for data-rate reduction techniques. Slightly higher reduction factors can be achieved with the Karhunen-Loeve transform. However, the requirements in terms of hardware and software are incomparably higher. The choice of DCT as an almost ideal

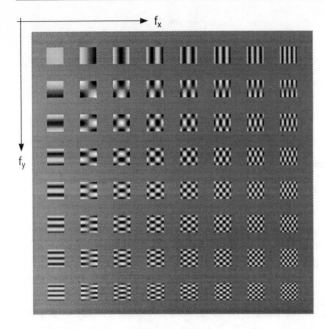

Fig. 4.4. DCT basis functions

transform for the JPEG standard and also for the MPEG standards was a trade-off between cost and benefit. The same applies to the fixing of the block size at 8×8 pixels. Larger blocks would lead to a minimally higher effectiveness of the data reduction, but the computation requirement increases geometrically in relation to the side length of the block. An early implementation of the algorithm in reasonably priced ICs could only be ensured with the 8×8 DCT.

4.1.3 Quantisation

Following the transformation into the frequency domain, the quantisation of the coefficients $G(f_x, f_y)$ can be carried out in accordance with the equation

$$G_Q(f_x, f_y) = \text{round}\left(\frac{G(f_x, f_y)}{Q(f_x, f_y)}\right) \tag{4.2}$$

with: f_x, f_y = spatial frequencies
     $G_Q(f_x, f_y)$ = quantised DCT coefficients
     $G(f_x, f_y)$ = unquantised DCT coefficients
     $Q(f_x, f_y)$ = quantisation step size

in which the quantised coefficients $G_Q(f_x, f_y)$ are computed by being divided by the required quantisation step size $Q(f_x, f_y)$ and then rounded to integers.

f_x →

f_y

16	11	10	16	24	40	51	61
12	12	14	19	26	58	60	55
14	13	16	24	40	57	69	56
14	17	22	29	51	87	80	62
18	22	37	56	68	109	103	77
24	35	55	64	81	104	113	92
49	64	78	87	103	121	120	101
72	92	95	98	112	100	103	99

$Q (f_x, f_y)$, luminance

f_x →

f_y

17	18	24	47	99	99	99	99
18	21	26	66	99	99	99	99
24	26	56	99	99	99	99	99
47	66	99	99	99	99	99	99
99	99	99	99	99	99	99	99
99	99	99	99	99	99	99	99
99	99	99	99	99	99	99	99
99	99	99	99	99	99	99	99

$Q (f_x, f_y)$, chrominance

Fig. 4.5. Examples of possible quantisation tables

This results in a linear quantisation characteristic. The denominator is generally different for each spatial frequency (f_x, f_y) and therefore an adaptation to the dynamic range of the human visual system can be achieved.

Figure 4.5 shows one table each of quantisation step size $Q(f_x, f_y)$ for luminance and for chrominance, which were determined by means of a psychophysiological experiment [LOHSCH]. In order to determine the values for the luminance the basis functions of the DCT are mixed with a background with an average grey-scale value. By operating a manual adjustment the subjects reduce the amplitude of the basis functions just so far as to be no longer perceptible. The detected 'just noticeable difference', the smallest identifiable difference, is used as the corresponding quantisation step size.

As can be seen, the quantisation tends towards being coarser the higher the spatial frequency. This is due to the fact that the human eye can perceive fine details with minimal dynamic content. Thus, in the case of luminance, the 11-bit amplitude resolution of the direct-current component (DC coefficient) $G(0,0)$ resulting from the DCT is limited to 7 bits – and hence to 128 amplitude steps for $G_Q(0,0)$, – as a consequence of its being divided by 16, whereas G_Q $(7,7)$ can only assume $2^{11}/99 = 21$ discrete values.

The quantisation tables shown in figure 4.5 are mentioned in the informative annexe to the JPEG standard, but are not stipulated for the compression. After all, they only represent the purely empirically determined optimum for the quantisation of luminance and chrominance of typical natural images of the sampling format of ITU-R BT.601 [ITU 601] with 720×576 pixels, when viewed at a distance of four times the image height. The tables actually used must be transmitted in the compressed data stream in order to enable the decoder to invert the quantisation.

To illustrate the interworking of DCT and quantisation, figure 4.6 shows an example taken from [WALLACE]. Numerical block *a* shows the sampling values of the original image and *b* shows the resulting DCT coefficients. Following quantisation with the aid of block *c*, the quantised coefficients in *d* are

139	144	149	153	155	155	155	155
144	151	153	156	159	156	156	156
150	155	160	163	158	156	156	156
159	161	162	160	160	159	159	159
159	160	161	162	162	155	155	155
161	161	161	161	160	157	157	157
162	162	161	163	162	157	157	157
162	162	161	161	163	158	158	158

a samples of the original

235,6	-1,0	-12,1	-5,2	2,1	-1,7	-2,7	-1,3
-22,6	-17,5	-6,2	-3,2	-2,9	-0,1	0,4	-1,2
-10,9	-9,3	-1,6	1,5	0,2	-0,9	-0,6	-0,1
-7,1	-1,9	0,2	1,5	0,9	-0,1	0,0	0,3
-0,6	-0,8	1,5	1,6	-0,1	-0,7	0,6	1,3
-1,8	-0,2	1,6	-0,3	-0,8	1,5	1,0	-1,0
-1,3	-0,4	-0,3	-1,5	-0,5	1,7	1,1	-0,8
-2,6	1,6	-3,8	-1,8	1,9	1,2	-0,6	-0,4

b DCT coefficients

16	11	10	16	24	40	51	61
12	12	14	19	26	58	60	55
14	13	16	24	40	57	69	56
14	17	22	29	51	87	80	62
18	22	37	56	68	109	103	77
24	35	55	64	81	104	113	92
49	64	78	87	103	121	120	101
72	92	95	98	112	100	103	99

c quantisation table

15	0	-1	0	0	0	0	0
-2	-1	0	0	0	0	0	0
-1	-1	0	0	0	0	0	0
0	0	0	0	0	0	0	0
0	0	0	0	0	0	0	0
0	0	0	0	0	0	0	0
0	0	0	0	0	0	0	0
0	0	0	0	0	0	0	0

d quantised coefficients

240	0	-10	0	0	0	0	0
-24	-12	0	0	0	0	0	0
-14	-13	0	0	0	0	0	0
0	0	0	0	0	0	0	0
0	0	0	0	0	0	0	0
0	0	0	0	0	0	0	0
0	0	0	0	0	0	0	0
0	0	0	0	0	0	0	0

e coefficients after inverse quantisation

144	146	149	152	154	156	156	156
148	150	152	154	156	156	156	156
155	156	157	158	158	157	156	155
160	161	161	162	161	160	157	155
163	163	164	163	162	160	158	156
163	164	164	164	162	160	158	157
160	161	162	162	161	160	159	158
158	159	161	161	162	161	159	158

f reconstructed samples

Fig. 4.6. An example of DCT and quantisation

subjected to the redundancy reduction as described in the following section. This completely reversible process has been omitted in the example for reasons of clarity. Following the inverse quantisation e and the inverse DCT, the reconstructed image f is obtained, which differs from the original to such a small extent that the difference cannot be detected by the viewer. Without accepting these invisible differences, i.e. by performing a so-called lossless coding, it would only be possible, on average, to achieve a data-rate reduction by a factor of two or less.

4.1.4 Redundancy Reduction

Following the quantisation of the DCT coefficients there is a redundancy reduction which further reduces the amount of data for the representation of the corresponding block. A prerequisite for this is the rearranging of the coefficients in zigzag order, as shown in figure 4.7. The two-dimensional matrix is mapped into a one-dimensional field such that the coefficients are ordered from the lowest frequency to the highest. This typically causes the larger coefficients to appear at the beginning of the field, while the small coefficients, as well as many consecutive zeros, appear at the end. This circumstance, which is illustrated in the "pizza" example in figure 4.7, is very convenient for the subsequent combination of run-length coding and Huffman coding.

As shown in figure 4.8, the alternating-current (AC) coefficients are coded such that the number of consecutive zero coefficients and the next coefficient not equal to zero are combined to form a pair of numbers. From the sequence 0,0,–2 results the pairing (2,–2), from 0,0,0,0,–1 the pairing (4,–1). Finally,

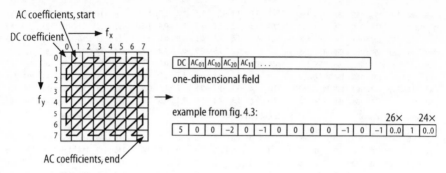

Fig. 4.7. Reordering of the coefficients in zigzag order

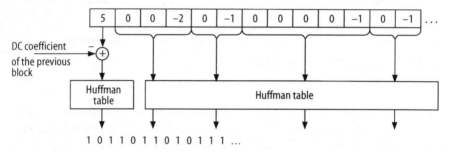

Fig. 4.8. Redundancy reduction

simply by referring to a coding table, one will find the most frequent pairs of numbers represented by few bits, while more bits will represent less frequent pairings. This method of coding, after Huffman, makes for a minimisation of the amount of data.

As the DC coefficient typically carries the greatest share of energy of the block (and is therefore only seldom equal to zero), it is subjected to a special treatment. By subtraction from the DC coefficient of the previous block its value will, on average, be considerably reduced. The resulting difference value is then also subjected to a correspondingly adapted Huffman coding.

The bit stream generated in this way, together with some signalling information such as quantisation tables and details concerning the resolution, etc., is then stored and transmitted as the JPEG bit stream. The algorithm described above typically entails a bit-rate reduction factor of between four and eight, but for most of the image content, despite this reduction factor, the difference between the image reconstructed in the decoder and the original is subjectively almost indiscernible.

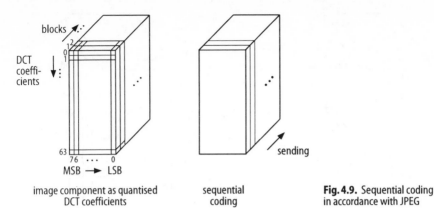

blocks

DCT coefficients

MSB → LSB

image component as quantised DCT coefficients

sequential coding

sending

Fig. 4.9. Sequential coding in accordance with JPEG

4.1.5 Specific Modes

The algorithm described is also referred to as "baseline JPEG". For most applications it constitutes a suitable method for data reduction. For some specific applications, however, specific modes are defined in the JPEG standard, which are described in the following.

Figure 4.9 shows the so-called sequential coding of quantised coefficients, which is identical to the already described baseline algorithm. The left side of the figure depicts the required amount of data as a rectangular prism on which the edge lengths are determined by the number of blocks per picture, the number of coefficients per block, and the number of bits required per coefficient. For convenience, it has been assumed that all coefficients are represented by 8 bits; in reality, of course, this number varies from coefficient to coefficient depending on the quantisation.

In sequential coding, the individual blocks are transmitted in sequence or are sequentially read out of a storage medium. The disadvantage of this procedure is that when data are transferred slowly (for example, during research in image data bases which are accessed via networks over considerable distances) the image is built up from top left to bottom right. Therefore it can take quite a long time for the viewer to receive a rough outline of the image and to decide whether or not it is the image sought. This costs the user of a data bank considerable time and unnecessarily wastes transmission capacity.

In this case help can be found in the so-called "progressive coding", which is shown in figure 4.10. The DC coefficients of all blocks are transmitted in a first run, and this is followed by the spectral selection in which the AC coefficients are transmitted in the order of increasing frequency. Thus, after the transmission of only a few bits, the viewer obtains a rough idea of the image as a whole, which at the start consists of 8×8 blocks of constant brightness and colour and which, through the inclusion of the other DCT basis functions, ap-

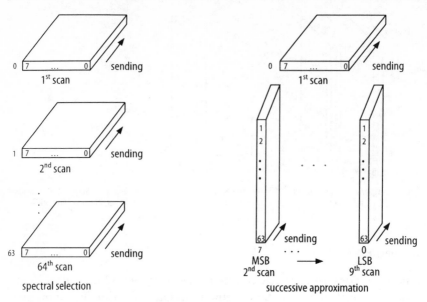

Fig. 4.10. Progressive coding in accordance with JPEG

pears in more and more detail in the course of the data transmission. Should the image turn out not to be the one required, this transmission can be broken off at any time; only the desired image need be completely transmitted. This saves time and transmission capacity.

A second example of progressive coding, the "successive approximation", can be seen in the right-hand side of figure 4.10. Following the transmission of the DC component, first the most significant bits (MSBs) of the AC coefficient, and then the least significant bits (LSBs) are transmitted. In this way an increasing image quality is obtained.

For some applications even the small differences between the reconstructed image and the original, as depicted, for example, in figure 4.6, are not acceptable. In medicine, for instance, the smallest distortion in an X-ray could influence a diagnosis. In order to supply the tools for such applications the JPEG standard offers a lossless mode. As shown in figure 4.11, this mode does not use transformation, as even the small rounding errors which occur inevi-

	selection value	prediction
	0	no prediction
C B	1	A
A X	2	B
	3	C
	4	A+B−C
	5	A+((B−C)/2)
	6	B+((A−C)/2)
X = actual pixel	7	(A+B)/2

Fig. 4.12. Lossless coding in accordance with JPEG, possible predictions

tably in the DCT-based mode of operation would cause a disturbance. Instead, based on the knowledge of the neighbouring pixels, a prediction of the actual pixel is made. The resulting difference from the actual value is coded with the aid of a Huffman table. Seven predictors, as listed in figure 4.12, are available for the prediction. In each case the most suitable prediction is chosen by the encoder and signalled to the decoder.

The data reduction factor obtained with lossless coding has an average value of <2 for natural images, while for the baseline algorithm, with almost the same subjective quality, an average factor of 4 to 8 is achieved. For this reason the lossless mode can only be considered as a solution for special applications. For example, for X-rays which have a high proportion of black areas the reduction factor obtained is much higher than two.

A further possibility of achieving a gradual increase in the resolution of an image is offered by the "hierarchical coding" which can be paired with all other JPEG modes. The relevant block diagram is shown in figure 4.13. The image to be coded is subsampled horizontally and vertically with the same factor, which must be a power of two (2^N). The resulting image is coded with a non-hierarchical JPEG encoder and transmitted. At the same time it is reconstructed at the decoder and oversampled by a factor of two so that it is avail-

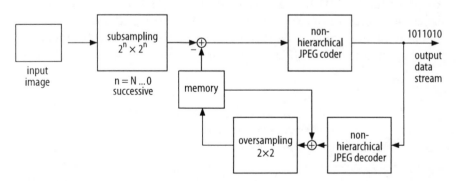

Fig. 4.13. Hierarchical coding in accordance with JPEG

able as a prediction for the next coding step, which uses the image now only subsampled by the factor 2^{N-1}. This continues until the best available resolution of the image (subsampling factor $2^0 = 1$) has been transmitted. In this way a successively increasing resolution is obtained during transmission, which can also be used in connection with lossless coding.

Concerning the use of the JPEG standard for transmitting moving pictures, none of these special coding modes is of any significance; firstly, because the transmission must take place in real time anyway and therefore a successively increasing resolution of the image quality would not be appropriate and, secondly, because with lossless coding a satisfactory data-rate reduction cannot be achieved. As mentioned before, M-JPEG was not standardised, which signifies that it is an adaptation of the baseline algorithm to moving pictures. If the term "JPEG" is used, then what is typically meant is baseline JPEG.

4.1.6 Interchange Format

In order to ensure compatibility and interchangeability between application environments, the JPEG standard defines an interchange format which on the one hand determines the structure of the bit stream, and on the other hand also determines the subsampling matrix. Figure 4.14 shows the relationship between the various components (with different resolutions) of the input image. The samples of the highest-resolution component (usually the luminance Y) are arranged in an orthogonal matrix. In the case of lower-resolution components (for example, chrominance signals C_B and C_R) the subsampling factor must be horizontally and vertically equal and the location of the samples must be midway between those of the higher-resolution component. The same is true for lower-resolution components, as can occur in specialised applications.

This definition is in accordance with the usual format in computer applications like, for example, Postscript Level 2, or Apple Quicktime, but not with Recommendation ITU-R BT.601 which relates to television. As stated by ITU, the chrominance for the 4:2:2 format is only subsampled horizontally and the chrominance sampling matrix is otherwise co-located with the luminance sampling matrix. If, in M-JPEG applications, one were to adhere to the ITU

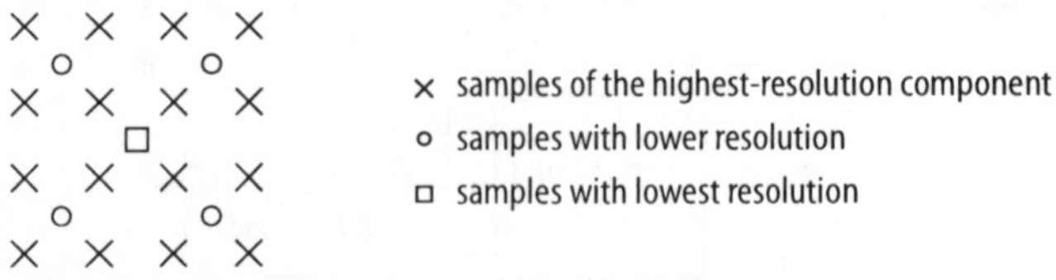

× samples of the highest-resolution component
○ samples with lower resolution
□ samples with lowest resolution

Fig. 4.14. Interchange format in accordance with JPEG

recommendations as well as to the JPEG format, a costly format conversion entailing quality losses would be required before the encoder and after the decoder. Therefore, most applications operating with the JPEG algorithm for the transmission and storage of moving pictures adhere to the ITU recommendation and infringe the JPEG format. Regarding such applications, this procedure should not cause any problems as long as data are not exchanged among codecs from different manufacturers.

4.2 Coding in Accordance with the MPEG Standards

The JPEG standard was originally conceived for still pictures and is therefore not completely suitable for data reduction of moving pictures. Firstly, the lack of a standard for M-JPEG means that there are considerable incompatibilities between the methods used by different manufacturers. Secondly, JPEG does not take advantage of the similarity of successive moving pictures, and therefore the reduction factor is limited to unnecessarily low values. Thirdly, JPEG neither provides for a coding of the associated audio information nor for a multiplexing of video and audio signals.

For this reason the Moving Pictures Experts Group defined an algorithm for coding moving pictures – including associated audio – which does not have the above shortcomings. The first step in this direction was the development of the MPEG-1 standard which was designed for use with computers and multimedia, particularly for the storage of video on CDs with a data rate of 1.15 Mbit/s. In order to obtain an acceptable image quality at this very low data rate it was necessary to include a number of limitations in the standard, e.g. that the resolution of the image to be coded must not be higher than 352×288 pixels, that the frame rate must not exceed 30 Hz, and that only progressive image sampling must be supported. An upper limit of 1.5 Mbit/s was set for the data rate. Adherence to this so-called 'constrained parameter set' ensures that any MPEG-1 decoder can decode the data. Owing to the early availability of suitable components, it was only a short time after the completion of the standard that applications came on the market which went beyond the established parameters and made it possible to also code images in the ITU-R BT.601 format, with 720×576 pixels (often referred to as MPEG-1+, MPEG-1.5 or similar). However, since the treatment of interlaced scanning is not specified in the MPEG-1 standard, applications using this type of scanning must be seen as individual solutions, all of which became obsolete with the completion of the MPEG-2 standard in November 1994.

MPEG-2 includes an extension of the MPEG-1 standard for the transmission of television signals, therefore it takes interlaced scanning into consideration and offers a succession of quality steps and options which are usu-

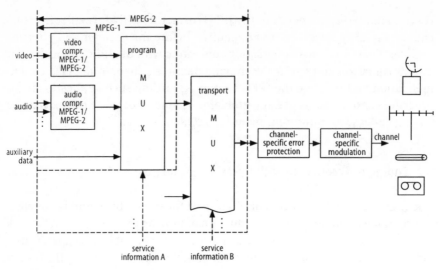

Fig. 4.15. Overview of MPEG-1 and MPEG-2 standardisation ranges

ally represented in the profile/level table, which is described later. The Main Profile at Main Level (MPML), which is intended for the transmission of standard television signals and allows for data rates of up to 15 Mbit/s, has shown in MPEG tests that an image quality comparable to PAL can be achieved even at bit rates of about 6 Mbit/s and that, at a code rate of 9 Mbit/s, the reconstructed image is almost indistinguishable from that of the original (so-called visual transparency). Originally, a third level, MPEG-3, was planned for the MPEG algorithm, to allow the coding of HDTV signals. However, as the coding of high-definition images has been incorporated in the profile/level table of MPEG-2 there is no requirement for MPEG-3.

The work of the MPEG group nevertheless continued in the direction of higher data-compression factors or extremely low data rates, e.g. below 64 kbit/s, and was subsumed under the name of MPEG-4. This "very low bit-rate coding", which might be applied in video conferencing or mobile services, will no longer exclusively rely on the DCT-based technique, as used in MPEG-1 and MPEG-2, but, with the aid of object-oriented coding, tries to achieve much higher reduction factors, albeit at a lower-grade quality. The use of MPEG-4 in broadcast services is currently not being envisaged and therefore MPEG-4 will not be discussed here.

Figure 4.15 gives an overview of the MPEG-1 and MPEG-2 standardisation range. In the MPEG standards not only the video-data reduction is defined, but also the audio-data reduction which was discussed in detail in chapter 3. It includes, moreover, the multiplexing of video, audio and other information, such as videotext, which are to be transmitted in a communal bit stream.

MPEG-1 only offers the multiplexing of one single programme, which is distinguished by the fact that all components (video, audio and auxiliary data) have one common time base. With MPEG-2, on the other hand, several programmes can be combined to form one common data stream. All other transmission parts outside the dashed lines representing the various blocks, i.e. the error protection adapted to the channel characteristics, the modulation used for the transmission, and the service information (see chapter 5), are not included in the MPEG standards. In Europe these elements are the responsibility of the European Digital Video Broadcasting (DVB) Project and have already been defined.

In the next section, the algorithm, i.e. the principle which is used by MPEG-1 and MPEG-2 for video coding, will be explained, followed by a discussion of the differences between the two standards.

4.2.1 Block Diagrams of Encoder and Decoder

The MPEG standards, too, make use of the discrete cosine transformation for spatial decorrelation. In addition, however, they also exploit the similarity of successive pictures and therefore, with no change in the quality, achieve considerably higher compression factors. Figure 4.16 shows the principle of differential coding, on which the data-rate reduction in the temporal direction is based. In the encoder the decoded signal, which is delayed by time factor τ, is subtracted from the actual signal in order to decorrelate the distribution of the amplitude values. The decoder inverts this step. τ can represent the duration of a pixel (in this case the neighbouring pixels are subtracted in the encoder), the duration of a line (corresponding to the subtraction of pixels in neighbouring lines), or the duration of an image (corresponding to the subtraction of equally located pixels in a succession of pictures). The last case has been put into effect in MPEG, as the similarities between the images are exploited by the DCT.

As can be seen, there is a complete decoder in the feedback branch of the encoder so that exactly *that* signal is used for the subtraction which is later generated in the decoder and added to the incoming signal. In this way the coding process can work completely without loss (redundancy reduction) as long as the range of numbers has not been limited by way of rounding errors

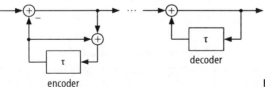

encoder   decoder

Fig. 4.16. Principle of differential coding

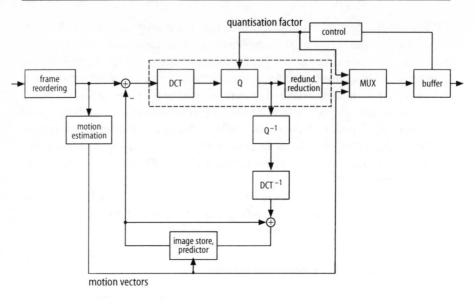

Fig. 4.17. Block diagram of an MPEG video encoder

or quantisation. If the input signal is quantised with 8 bits and thus assumes values in the range of 0 to 255, then values between –255 and +255 (9 bits) can occur at the output of the encoder in the most extreme case. Simple methods, which achieve a data reduction solely by performing the differential coding, limit the difference signal in its word length to, for instance, 6 bits (range of values –32 … + 31), but accept artefacts (slope overloads) on all steep signal transitions (so-called differential pulse code modulation [DPCM]). For MPEG the difference values are fed into the subsequent DCT with an unchanged 9-bit word length. The gain in terms of data reduction results from the decorrelation.

Figure 4.17 shows the block diagram of an MPEG video encoder, the components of which will be discussed in detail in the following paragraphs. The steps DCT, quantisation and redundancy reduction that can be seen in the centre, framed by dashed lines, are also found in JPEG. In addition to this, the encoder features differential coding with an image store for time delays by the value of τ. This prediction is supported by motion estimation which seeks the best possible match in the previous image for the image to be coded in each block. As the decoder requires the motion information in order to invert the prediction, the calculated motion vectors are also transmitted in the bit stream. In order to also exploit the similarity with successive pictures, the coding is preceded by a reordering of the sequence of pictures.

The second important new functionality included in MPEG-2, as compared to JPEG, is the control of the quantiser to regulate the transmission of a

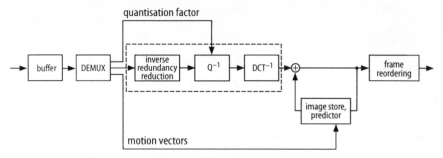

Fig. 4.18. Block diagram of an MPEG video decoder

constant data rate. At the output of the encoder there is a buffer which absorbs the incoming data of varying rates and passes them on at a constant data rate. If there is any danger of the buffer overflowing, the buffer will cause the quantisation factor (see section 4.2.4) to produce a coarser quantisation so that less data are transferred to the buffer. A constant data rate ensues as a consequence.

The corresponding decoder is shown in figure 4.18. The data which arrive at a constant data rate are absorbed by the input buffer and transferred at varying rates to the demultiplexer which separates the coded image data from the required auxiliary information (especially the quantisation factor and the motion vectors). The inverse quantisation follows the inverse redundancy reduction which evaluates the quantisation factor transmitted in the bit stream. The inverse DCT transforms the coefficients back into the spatial domain whereupon the predicted values are added. For this step the motion vectors are required which were calculated by the encoder and transmitted in the data stream. Finally the decoded images are put back in the right order.

It is obvious from the block diagrams that the construction of the decoder is considerably simpler than that of the encoder. In particular no motion vectors have to be defined, as these are generated by the encoder. The motion estimation is, however, the step which requires the most computing power and this is why it will be some time before a single-chip MPEG encoder is manufactured, whereas decoder ICs are already available.

4.2.2 Motion Estimation

To illustrate the motion-estimation-supported prediction an example is given in figure 4.19. This shows a ball which has moved minimally to the right and downward from image 1 to image 2. For coding purposes the images were broken down into so-called macroblocks with 16×16 pixels (corresponding to 2×2 DCT coding blocks). To code the thickly drawn macroblock in image 2,

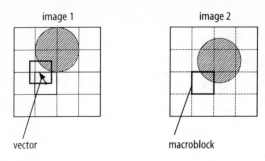

image 1   image 2

vector   macroblock   **Fig. 4.19.** Example of motion estimation

the difference between the corresponding macroblocks in image 2 and image 1 is computed. There remains in the top right-hand corner of the resulting macroblock a relatively large difference value because the ball has not yet appeared in the corresponding macroblock in image 1. Accordingly, a relatively large number of bits is required for coding. A considerably better prognosis can be made by computing the difference from a macroblock which has been shifted in the right direction, as shown in figure 4.19. However, in this case the calculated motion vector must also be transmitted so as to enable the decoder to invert the process.

The selection of macroblocks as the basic element for motion estimation offers the advantage that the same vector can be used for the luminance components as for the chrominance components. If the chrominance is subsampled horizontally and vertically by the factor 2 (4:2:0 coding, see section 4.1.6 and 4.2.5), one macroblock is made up of four luminance blocks as well as of one block each of the two colour difference signals, as shown in figure 4.20. A corresponding definition holds for formats 4:2:2 (as in ITU-R BT.601) and 4:4:4 (no subsampling), which are provided only by MPEG-2.

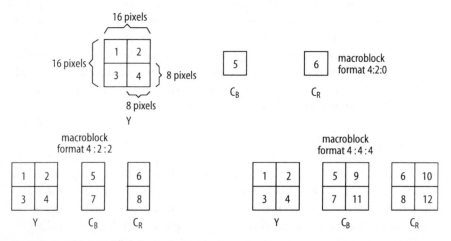

Fig. 4.20. Possible macroblock structures

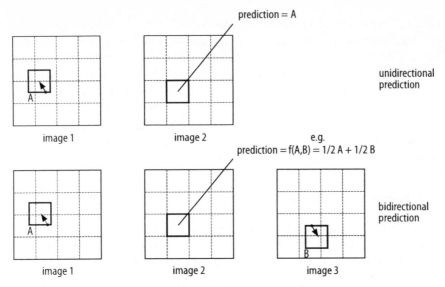

Fig. 4.21. Unidirectional and bidirectional prediction

The MPEG standards do not specify which algorithm is to be used to calculate the motion vectors. Usually a block-matching technique is used [MUSMANN], by which, for example, the macroblock is shifted to all possible positions within a given search area, and the position for coding is chosen where the difference to the actual macroblock is the smallest. In "full-search block matching" all positions are actually checked in steps of half a pixel (realisation by interpolation), whereas hierarchical procedures commence with larger steps, locating a provisional minimum, and then continue with smaller steps in a search area around the provisional minimum. In any case, applying the search algorithm requires a very large computational effort, which means that for hardware realisations the search area is often considerably reduced. However, the size of the search area plays a decisive role in determining the quality of the reconstructed picture. For example, if the search area is restricted to ±10 pixels horizontally and ±5 pixels vertically, then, if wide-range shifting of more than 10 pixels occurs horizontally from image to image (e.g. due to a fast camera swing) the prediction fails. The reduced coding efficiency resulting from this causes a considerable quality loss.

Prediction using future images has also been included in MPEG in order to further increase the coding efficiency. Figure 4.21 shows what is to be understood by "bidirectional prediction". As opposed to unidirectional prediction, the macroblock in image 3 which follows image 2, the image which is to be coded, is additionally searched. The prediction can thus, for example, result from the mean value between macroblock A and macroblock B. If image 1 or image 2 does not yield a usable prediction (which could be the case, for example, in cut images), the encoder can decide that only the other prediction must be used. Both orientations of the prediction could fail in the case of fast movements. In such a case the encoder is free to revoke the prediction and save the bits for coding the vectors. Even though bidirectional prediction means doubling the expense, it was included in the standard because it considerably increases the coding efficiency. A bidirectionally predicted macroblock can on average be coded with only about half the amount of data required for a unidirectionally predicted macroblock.

4.2.3 Reordering of Pictures

Bidirectional prediction not only doubles the computational effort of the motion vector calculation but also necessitates the already-mentioned reordering of the pictures. The upper part of figure 4.22 shows an extract from a frame sequence. If one were to use a prediction for all pictures, it would not be possible to find a starting point for the decoding, as each picture is dependent on the one before and after it. For this reason one picture, at suitable intervals, is coded without the help of prediction (so-called I-pictures, intraframe-coded). If an I-picture is inserted every 12 frames, then, when the television receiver is switched on or the programme changed, it takes a maximum of half a second before the decoding can begin.

Between the I-pictures, so-called P-pictures, which have been unidirectionally predicted from the preceding I-picture or P-picture, are inserted as supporting points. The interleaved bidirectionally predicted pictures (B-

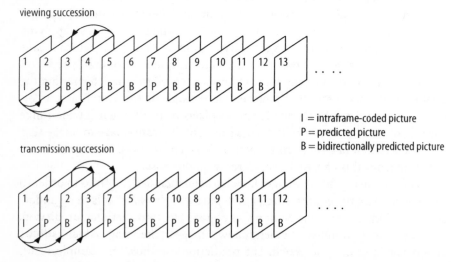

Fig. 4.22. Reordering of pictures

pictures) result from the I-pictures or P-pictures which appear before and after. As can be seen from the interdependencies of the pictures, marked by arrows, the decoder must always recognise the subsequent P-picture or I-picture before the interleaved B-pictures can be decoded. The same applies to the encoder since there is a complete decoder in its feedback branch. That is why the sequence of the pictures is changed before transmission. The framing I-pictures and P-pictures are always transmitted before the interleaved B-pictures. Therefore, the encoder has to have a total of four image stores and to carry out the following steps for two consecutive B-pictures: coding the I-picture and retaining the original for the prediction; merely retaining the originals of the B-pictures for the time being; coding the P-picture and retaining the original for the prediction of B-pictures; then coding the B-pictures. The decoder, by comparison, requires only two image stores, irrespective of the number of consecutive B-pictures to be decoded, as it only needs to retain the framing I-pictures and P-pictures; the B-pictures can be displayed directly after decoding and removed from storage. The decoding of the B-pictures requires extra storage space, the extent of which depends on the implementation.

The determination of the sequence of the I-pictures, P-pictures and B-pictures (so-called group-of-pictures structure) depends solely on the decision of the encoder. The decoder can process all sequences since it never requires more than just a little over two image stores for the decoding. The complexity of the encoder as well as the image quality achieved depend essentially on the structure of the group of pictures. The structure shown in figure 4.22 has proved to be a viable trade-off between cost and performance; however, this structure is by no means mandatory.

4.2.4 Data-rate Control

As already mentioned, a constant data rate is achieved by controlling the quantiser. If there is a danger of the output buffer of the encoder overflowing – for instance, because a picture content with fast movement and thus with a lot of high frequencies has to be coded or because the prediction malfunctions on account of the limitation of the search area – the quantisation step is increased, resulting in a coarser quantisation and possibly in a visible deterioration of the picture quality. When the situation eases, the quantisation step can again be decreased. It is on the quantisation that the amount of data input to the buffer depends. For a minimum quality of the reconstructed picture to be guaranteed, the data-rate control does not affect the quantisation of the DC component of intraframe-coded pictures. The DC coefficient is always divided by eight. On the other hand, the DC coefficients for P-pictures and B-pictures, just as the AC coefficients, are quantised with a step size taken from a similar table to the one used for JPEG. This, in addition, is followed by a further quantisation which performs the data-rate control.

DC coefficient for I-pictures:

$$G_Q(0,0) = \text{round}\left(\frac{G(0,0)}{8}\right) \tag{4.3}$$

Otherwise:

$$A(f_x, f_y) = \text{round}\left(\frac{G(f_x, f_y)}{Q(f_x, f_y)}\right)$$

$$G_Q(f_x, f_y) = \text{round}\left(\frac{A(f_x, f_y)}{Q_F/8}\right) \tag{4.4}$$

with:
f_x, f_y = spatial frequencies
$G_Q(f_x, f_y)$ = quantised DCT coefficients
$G(f_x, f_y)$ = unquantised DCT coefficients
$A(f_x, f_y)$ = auxiliary quantity
$Q(f_x, f_y)$ = entries in the quantisation table
Q_F = quantisation factor

Here Q_F is the quantisation factor for the data-rate control, which varies from macroblock to macroblock and which can be set by the buffer control to a value between 1 and 31. By this means entries in the quantisation table are variegated by a common factor which can be set in steps of 1/8. If the setting $Q_F = 31$ does not suffice to prevent a buffer overflow, so-called "skipped macroblocks" occur, i.e. macroblocks which are coded with 0 bits and are therefore skipped. In this case the decoder sets all motion vectors and all coefficients to zero, resulting in the macroblock of the previous image simply being repeated. Of course such macroblocks are generally very conspicuous as they do not fit exactly. In contrast, if the image content is so simple that even at $Q_F = 1$ too few data are produced, then an emptying of the buffer is avoided by inserting so-called stuffing bits.

Contrary to the JPEG standard, quantisation tables were agreed for MPEG-1 and MPEG-2, and these can be seen in figure 4.23. In this way the bits for signalling these tables can be saved. However, should the encoder discover that variant quantisation tables would lead to a higher coding efficiency, then it has the option to transmit those tables in addition to the others.

4.2.5 Special Features of MPEG-1

MPEG-1 was designed for use with computers and multimedia and particularly for the recording of videos on conventional CDs. To keep within the given limited data rate, the resolution of the image to be coded had to be re-

f_x							
8	16	19	22	26	27	29	34
16	16	22	24	27	29	34	37
19	22	26	27	29	34	34	38
22	22	26	27	29	34	37	40
22	26	27	29	32	35	40	48
26	27	29	32	35	40	48	58
26	27	29	34	38	46	56	69
27	29	35	38	46	56	69	83

$Q(f_x, f_y)$ for I-pictures

f_x							
16	16	16	16	16	16	16	16
16	16	16	16	16	16	16	16
16	16	16	16	16	16	16	16
16	16	16	16	16	16	16	16
16	16	16	16	16	16	16	16
16	16	16	16	16	16	16	16
16	16	16	16	16	16	16	16
16	16	16	16	16	16	16	16

$Q(f_x, f_y)$ for P- and B-pictures

Fig. 4.23. Standard quantisation tables in MPEG-1 and MPEG-2

stricted to the so-called "source input format" (SIF) (see sections 2.2 and 4.1). This is to be understood as essentially half the resolution compared to an image conforming to Recommendation ITU-R BT.601 in conjunction with the 4 : 2 : 0 colour representation. In figure 4.24 the sampling structures of the two formats are compared. Whereas the chrominance in ITU-R BT.601 is only subsampled horizontally by a factor of two and whereas apart from that the sampling grid coincides with that of the luminance, the chrominance in the SIF is subsampled in both directions and its samples lie midway between those of the luminance (as with the JPEG interchange format, see section 4.1.6).

To transmit or store a television signal with MPEG-1 it is necessary to perform a format conversion before the encoder and after the decoder. This is illustrated by the transmission chain in figure 4.25, where "channel" can also be interpreted as a recording on CD. By way of an example the individual steps

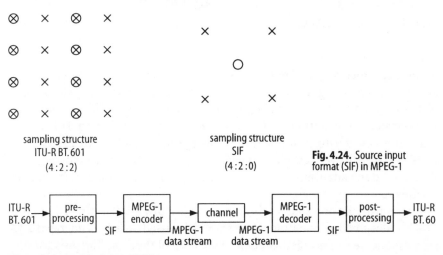

sampling structure
ITU-R BT.601
(4 : 2 : 2)

sampling structure
SIF
(4 : 2 : 0)

Fig. 4.24. Source input format (SIF) in MPEG-1

ITU-R BT.601 → pre-processing → SIF → MPEG-1 encoder → MPEG-1 data stream → channel → MPEG-1 data stream → MPEG-1 decoder → SIF → post-processing → ITU-R BT.60

Fig. 4.25. Processing chain in MPEG-1

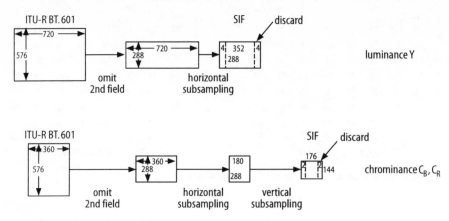

Fig. 4.26. Pre-processing in MPEG-1

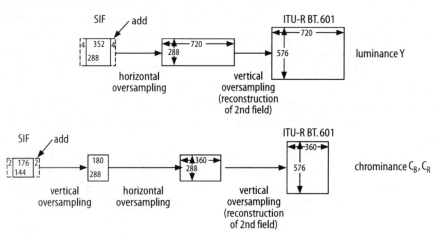

Fig. 4.27. Post-processing in MPEG-1

for the pre-processing and post-processing are shown in detail in figures 4.26 and 4.27. In changing the format to SIF, in the easiest case, one field is completely omitted for the time being. The use of a decimation filter at this stage would result in the motion phases of the two fields being blurred. This would cause artefacts during movement in the image. Should a filter be used at this stage to lessen the vertical aliasing which occurs by omitting a field, then this should be done adaptively in those image regions where movement is only minimal.

Afterwards, the horizontal subsampling is performed. This time, though, a suitably designed decimation filter should be used which, with regard to chrominance, ensures that the required position of the sampling points is observed. Some hardware implementations do not include filters at this point or elsewhere. They carry out the subsampling by simply omitting pixels. This, however, causes aliasing, which has a negative effect on the coding quality. For the chrominance components another vertical subsampling follows (here again, if possible, with the correct filter in terms of systems theory) with the object of achieving the 4 : 2 : 0 format. Since because of the macroblock structure the number of luminance pixels per line (just as the number of lines) must be divisible by 16 and the number of chrominance pixels per line must be divisible by 8, some pixels in the right and left margins of the picture will be discarded (delineated by broken lines in figure 4.26). The input signal for MPEG-1 coding, with a progressive sampling structure and a 25-Hz frame frequency, thus has a resolution of 352×288 luminance pixels and of 176×144 chrominance pixels.

Following the decoding process, the corresponding inverse steps must be carried out to reconstruct the original format. Horizontal oversampling takes place subsequent to the addition of black pixels to the left and right margins of the picture and subsequent to vertical oversampling of the chrominance components. The reconstruction of the second field can also be performed by oversampling; however, the second motion phase which was eliminated by the encoding can, of course, not be reconstructed. This would be possible by means of costly motion-adaptive processing, but even then only partially.

4.2.6 Special Features of MPEG-2

The considerable limitations of the MPEG-1 standard do not recur in MPEG-2. MPEG-2 provides for the coding of signals in accordance with ITU-R BT.601 as well as for the use of the 4 : 2 : 2 chrominance format and even of the 4 : 4 : 4 format. The main profile proposed by the European DVB Project for television broadcasting utilises the 4 : 2 : 0 format which, however, differs from the 4 : 2 : 0 definition in MPEG-1 in the position of the chrominance sampling grid in relation to that of the luminance. Apart from this, the interlaced scanning is taken into account. The left part of figure 4.28 shows the 4 : 2 : 0 format with a progressive sampling grid, where the luminance sampling points are marked by crosses and the chrominance sampling points by circles. The grid differs from its counterpart in MPEG-1 and JPEG in that the vertical samples of the chrominance lie midway between the luminance lines but that horizontally they remain unchanged as compared to ITU-R BT.601. For MPEG-1 a horizontal filter has to be employed to enable the subsampling. So this same filter can be used for the horizontal phase shifting. Thus a format is created which conforms with the usual conventions in the field of computers. Right from the beginning, however, MPEG-2 was aimed at applications in television engineering, where a sampling structure like that depicted in figure 4.28 is used and where a horizontal shifting of the sampling grid would mean

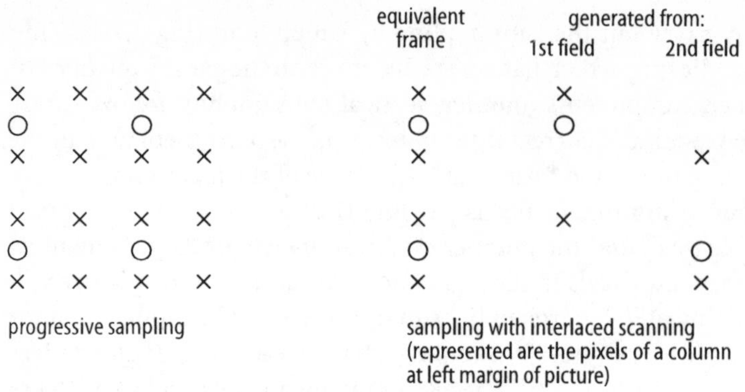

				equivalent frame	generated from: 1st field	2nd field
×	×	×	×	×	×	
○		○		○	○	
×	×	×	×	×		×
×	×	×	×	×	×	
○		○		○		○
×	×	×	×	×		×
progressive sampling				sampling with interlaced scanning (represented are the pixels of a column at left margin of picture)		

Fig. 4.28. Sampling structure 4 : 2 : 0 in MPEG-2

extra costs. For sampling using interlaced scanning the position of the samples does not change; the assigning of the chrominance lines to the fields is indicated in the right half of figure 4.28, which shows the first column of an image as part of a frame and as part of the two fields.

The high profile of the MPEG-2 standard also permits the 4:2:2 format, which is shown in figure 4.29, whereas the 4:4:4 format, as depicted in figure 4.30, has been provided for but is not contained in the profiles up to now. For future applications the standard can be supplemented by further profiles with the option of a 4:4:4 format. The 4:2:2 format conforms to ITU-R BT.601, whereas for the 4:4:4 representation the luminance and chrominance grids are congruent and superimposed on each other.

The interlaced scanning also affects the performance of the DCT. In MPEG-2 the encoder can choose between the field mode and the frame mode, as shown in figure 4.31. The two fields of the picture are represented by black and white picture lines. In the frame mode each macroblock is divided into

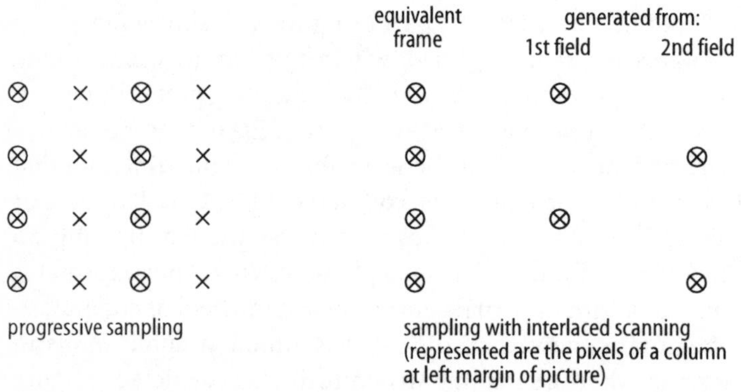

				equivalent frame	generated from: 1st field	2nd field
⊗	×	⊗	×	⊗	⊗	
⊗	×	⊗	×	⊗		⊗
⊗	×	⊗	×	⊗	⊗	
⊗	×	⊗	×	⊗		⊗
progressive sampling				sampling with interlaced scanning (represented are the pixels of a column at left margin of picture)		

Fig. 4.29. Sampling structure 4 : 2 : 2 in MPEG-2

		equivalent frame	generated from:	
			1st field	2nd field

⊗ ⊗ ⊗ ⊗   ⊗   ⊗

⊗ ⊗ ⊗ ⊗   ⊗   ⊗

⊗ ⊗ ⊗ ⊗   ⊗   ⊗

⊗ ⊗ ⊗ ⊗   ⊗   ⊗

progressive sampling   sampling with interlaced scanning
                                        (represented are the pixels of a column
                                        at left margin of picture)

Fig. 4.30. Sampling structure 4 : 4 : 4 in MPEG-2

8×8 DCT blocks in such a way that each 8×8 block contains both motion phases. For fast movements in the picture, however, this is a disadvantage since the breaking up of vertical edges in the image, caused by interlaced scanning, leads to high vertical frequencies and therefore to low coding efficiency. In this case it is more convenient to rearrange the fields as shown in

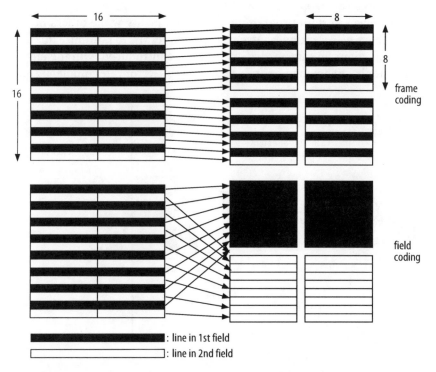

: line in 1st field
: line in 2nd field

Fig. 4.31. Possible distributions of 8×8 DCT blocks within a macroblock

the lower part of the representation. The chrominance results in the 4 : 2 : 0 format having two coding blocks of the size 8×4 instead of one coding block of the size 8×8. For a static image or an image with little movement the frame coding is generally more efficient as spatially neighbouring lines are usually more closely correlated. The decision as to which method is to be used is made by the encoder.

MPEG-2 differs from JPEG and MPEG-1 in that an optional non-linear quantisation of DCT coefficients is also possible. As a rule, this results in a higher coding efficiency. There is also the possibility of a modified run-length coding, which can be better adapted to the interlaced scanning using a different zigzag order and a second Huffman table.

A further innovation in the MPEG-2 standard are the different types of scalability, the meaning of which is explained below. The degradation of a digital television transmission is generally very abrupt. Even a small increase in the bit-error rate can cause a transition from perfect reception to total disruption. If the power budget of the transmission path does not include enough margin, e.g. if too small a receiving antenna is used for a satellite transmission, changeable weather conditions can cause a situation in which the picture continually vanishes and reappears. In the case of terrestrial transmissions it can happen that one household at the edge of a coverage area can still receive a certain programme in perfect quality while others in the next street have no reception whatsoever (cf. chapter 11). To minimise this unpleasant behaviour an intermediate stage can be introduced with the aid of scalability tools, as shown in figure 4.32. At a low bit-error rate completely un-

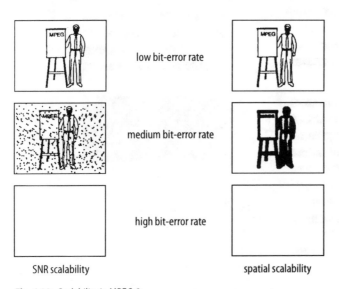

low bit-error rate

medium bit-error rate

high bit-error rate

SNR scalability

spatial scalability

Fig. 4.32. Scalability in MPEG-2

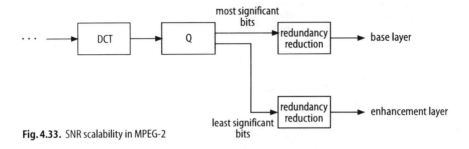

Fig. 4.33. SNR scalability in MPEG-2

disturbed reception is possible. If the bit-error rate increases to a higher value, then, in the case of SNR (signal-to-noise ratio) scalability, the picture received will be noisy but still acceptable. However, this is not noise as we understand it, but noise due to quantisation errors, combined with signal and block structures. By comparison, in the case of spatial scalability, at the increased bit-error rate, a lower-resolution image is received. It is only with even higher bit-error rates that the reception is completely disrupted. If it is envisaged to exploit the possibilities arising from the utilisation of scalability in source coding in a digital television system, "hierarchical modulation" (cf. section 11.5) will be required. The European DVB Project does not include either SNR scalability or spatial scalability in its system specifications.

Figure 4.33 shows the implementation of SNR scalability using a section of an MPEG-2 encoder. After quantisation, the DCT coefficients are divided into most significant and least significant bits and subjected to a separate redundancy reduction. In the channel encoder the so-called base layer with the most significant bits is then furnished with a better error protection for transmission than the least significant bits of the enhancement layers. Hence, in the case of a deterioration of the transmission characteristics of the channel it is the less important bits of the coefficients that can no longer be received, a fact which becomes noticeable as noise-like interference.

An encoder for spatial scalability is quite differently constructed (see figure 4.34). The image to be coded is first subsampled horizontally and vertically by a factor of two and then fed into an MPEG-2 encoder for this reduced resolution. The resulting base layer is then furnished with a high-grade error protection in the channel encoder. In the encoder the base layer is immediately decoded and, after reconversion to the original resolution, used as prediction for the enhancement layer, which is then furnished in the channel encoder with a lower-grade error protection. A particular advantage of spatial scalability is that it enables the compatible transmission of HDTV signals.

By combining both scalabilities it is possible, moreover, to achieve a multistage degradation. Two other types of scalability should be mentioned here:

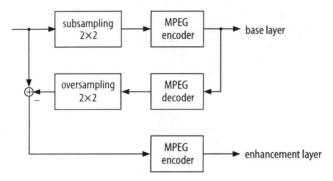

Fig. 4.34. Spatial scalability in MPEG-2

temporal scalability, which reduces the temporal resolution, i.e. the frame frequency, and data partitioning, similar in character to SNR scalability. Neither will be further discussed here.

The profile/level table for MPEG-2 mentioned at the beginning will now be examined more closely. Figure 4.35 shows its construction. The so-called profiles, under which certain syntactical elements and hence certain algorithmic peculiarities are subsumed, are arranged from left to right. The levels, limiting certain parameters like resolution and data rates to their corresponding maximum values, are arranged from bottom to top.

The upper limits for definition and data rate can be seen in the boxes. The numbers in brackets refer to the lower layer or layers of the scalable coding. The parameters for the cancelled boxes have not been defined.

levels	simple profile	main profile	SNR scalable profile	spatial scalable profile	high profile
high level		1920×1152 pixels 80 Mbit/s			1920×1152 pixels (960×576) 100(80.25) Mbit/s
high-1440 level		1440×1152 pixels 60 Mbit/s		1440×1152 pixels (720×576) 60(40.15) Mbit/s	1440×1152 pixels (720×576) 80(60.20) Mbit/s
main level	720×576 pixels 15 Mbit/s	720×576 pixels 15 Mbit/s	720×576 pixels 15(10) Mbit/s		720×576 pixels (352×288) 20(15.4) Mbit/s
low level		352×288 pixels 4 Mbit/s	352×288 pixels 4(3) Mbit/s		
profiles	(main profile, without B-pictures)	(4:2:0, no scalabilty)	(main profile, + SNR scalability)	(SNR profile, + spat. scalability)	(spatial profile, + 4:2:2 coding)

Fig. 4.35. Profiles and levels in MPEG-2

The low level has been conceived for the coding of television pictures with reduced definition, similar to the already-known SIF in MPEG-1 (LDTV: low-definition television). However, due to the specified upper limit of 4 Mbit/s for the data rate, a considerably better quality can be achieved with MPEG-2 than with MPEG-1. The main level is used for the coding of signals in the standard definition of contemporary television (SDTV: standard-definition television, comparable to PAL; EDTV: enhanced-definition television, comparable to ITU-R BT.601). For the coding of HDTV (high-definition television) two levels are available: the high-1440 level and the high level, the latter providing a number of pixels per line appropriate for the 16 : 9 aspect ratio. However, the transmission of 16 : 9 signals is possible at all levels (and with all profiles).

The simple profile and the main profile differ from each other only in that in the simple profile bidirectionally predicted pictures are not permissible. Therefore the encoder and decoder are not so costly; however, the loss in coding efficiency is so high that the simple profile most probably will not be used in practice. For the time being, scalability is not possible in either of the profiles, and the chrominance format has been fixed at 4 : 2 : 0. The SNR-scalable profile permits the utilisation of SNR scalability, while the spatially scalable profile is added for the spatial scalability. It is the high profile which additionally permits the 4 : 2 : 2 chrominance format.

The table as a whole is composed to indicate downward-compatibility, which means that a decoder which can be allocated to a particular box in the table must also be able to decode the data streams of all the profiles and levels situated to the left and below that box. Correspondingly, each MPEG-2 decoder must be able to decode all profiles and all levels of an MPEG-1 data stream. The standard also defines particular characteristics which do not appear in any profile/level combination, such as the 4 : 4 : 4 coding. Possible future extensions of the table will be reserved for these. The boxes which have been defined up to now, however, may not be changed. The rationale for defining this matrix of profiles and levels is to make available a practical subset from the multiple tools of the MPEG-2 standard for certain uses. This obviates the necessity to equip all decoders for all eventualities, thus accelerating the implementation of the algorithm in the respective hardware and saving costs.

4.3 Summary

Chapter 4 discusses the video coding in accordance with JPEG, MPEG-1 and MPEG-2. The JPEG standard, which uses the DCT together with subsequent quantisation and redundancy reduction, was primarily developed as a standard for compressing still pictures. However, due to the early availability of efficient and reasonably priced IC solutions it was soon used for coding moving pictures. This so-called Motion JPEG varies, though, from manufacturer to manufacturer, and as a consequence the various solutions are incompatible with one another. In the domain of video editing systems JPEG has secured itself an established place.

Right from the beginning MPEG-1 was conceived for lower-quality applications in computers/multimedia, and the standard is now firmly established in this domain. Because of the attraction of achievable cost reduction by means of data reduction this standard was used beyond its design objectives for the transmission of television signals with a higher quality and data rate before the completion of MPEG-2. Just like JPEG, MPEG-1 uses transformation coding, but supplements this by exploiting the temporal similarities in pictures and by controlling the constancy of the data rate.

Finally, MPEG-2 provides all the necessary tools for the coding of television signals of very different qualities, from SIF (similar to VHS) to HDTV, thus offering the Digital Video Broadcasting (DVB) Project a toolbox for video coding. DVB has compiled the "Implementation guidelines for the use of MPEG-2 systems, video and audio" [ETR 154] which incorporate the list of restrictions that need to be placed on MPEG-2 parameters in oder to ease the implementation of MPEG-2 in DVB. ETR 154 thus warrants the compatibility between equipment from different manufacturers. This has made DVB the first major application of MPEG-2 video coding.

Symbols in Chapter 4

A	signal value of a DPCM pixel
$A(f_x, f_y)$	auxiliary function in the spatial-frequency domain
B	signal value of a DPCM pixel; bidirectionally predicted picture
C	signal value of a DPCM pixel
$C(f_x, f_y)$	scaling constants in the spatial-frequency domain
C_B	digital colour difference signal: blue
C_R	digital colour difference signal: red
f_x	spatial frequency in x-direction
f_y	spatial frequency in y-direction
$G(f_x, f_y)$	spectral coefficients of DCT, transformed from $g(x,y)$
$G_Q(f_x, f_y)$	quantised spectral coefficients of DCT
$g(x,y)$	image signal in x,y spatial domain
I	intraframe-coded image
N, n	power of 2
P	predicted image (unidirectional)
p	multiplier (integer)
$Q(f_x, f_y)$	quantisation step size for DCT coefficients
Q_F	quantisation factor for data-rate control
round	mathematical rounding to integers
x	spatial co-ordinate
Y	digital luminance signal
y	spatial co-ordinate
τ	delay in the DPCM loop

5   MPEG-2 Systems and Multiplexing

Apart from the coding of audio and video signals (see chapters 3 and 4) the MPEG-2 standard [ISO 13818] also defines the multiplexing of audio, video and auxiliary data into one single bit stream, the so-called MPEG-2 System ([TEICHNER, KNOLL]). The joining together, however, is not the only task of the multiplex. It must also provide transmission capacity for information about the current programme or broadcast and about the transmission path, as well as other information required either for the technical servicing or as a navigational aid through the maze of programmes offered to the viewer. This information was defined by the DVB Project ([ETS 468, ETR 211]). Further functions covered by the MPEG-2 System are provisions for clock recovery in the decoder, the synchronisation of video and audio in order to retain the synchronism of the lip movements, and the provision of transmission capacity for conditional-access data (cf. chapter 8). These functions are described in the first part of the MPEG-2 standard entitled "Systems". They cannot be portrayed here in all their dimensions, but the following sections should serve to give an insight into what are sometimes very complex interconnections.

5.1 Differences between Programme Multiplex and Transport Multiplex

The operation of multiplexing is shown as a rough block diagram in figure 5.1. First of all the video, audio and auxiliary data are packetised, i.e. they are divided into relatively large units ("packets") and furnished with controlling information. It is only after this that the data streams are combined into a single one, in which process these so-called "packetised elementary streams" (PES) are divided into smaller packets which are then multiplexed. This step can lead either to a "program stream" (PS) with a single unified time basis or to a "transport stream" (TS) with the possibility of transmitting several different time bases and therefore several programmes in one channel. The main differences can be seen below:

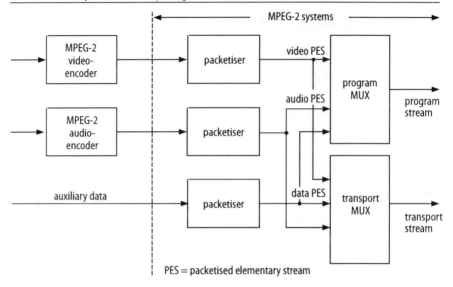

Fig. 5.1. Program multiplex and transport multiplex

Program multiplex:
- all elementary streams have one common time basis
- suitable for use in (relatively) error-free transmission channels (e.g. recording on hard disk)
- packets may be of variable length

Transport multiplex:
- several different time bases are possible
- suitable for use in error-prone channels (i.e. satellite transmission)
- fixed packet length of 188 bytes

The DVB Project has opted for the use of the transport multiplex for television broadcasting in Europe over satellite, cable and terrestrial transmitters, because this is the only one suitable for transmission on error-prone channels. The program multiplex will therefore not be discussed further.

5.2 Positioning of Systems in the ISO/OSI Layer Model

In order to ensure the applicability of the MPEG-2 system bit streams to most existing and future data networks, the Moving Pictures Experts Group has aligned the development of the multiplex with the ISO/OSI layer model ([ISO 7498], ISO = International Standardization Organization, OSI = Open Systems Interconnection). Figure 5.2 is an attempt to show the functionality of

Layer	Definition	Example: written message
7	application layer	reading and understanding
6	presentation layer	arrangement of writing, structuring
5	session layer	alphabet, language
4	transport layer	paper, ink
3	network layer	letterbox
2	data link layer	address on letter
1	physical layer	transportation by the mail company

Fig. 5.2. The ISO/OSI layer model

the model. The example has been kept as instructive as possible; therefore the analogy is not quite flawless.

The seven layers each perform different tasks during the transportation of a message. The seventh layer, the so-called application layer, can be associated with the recipient who reads and understands, as well as with the author who conceives and writes the message. The presentation layer is responsible for structuring the contents of the message. The coding of the text, i.e. the use of the agreed alphabet and the agreed language, takes place in the session layer. In layer four, paper and ink are the media used to make the message ready for despatch. The transition to layer three, i.e. the way to the letterbox, represents the transfer of the completed message to the transport network. The three lowest layers comprise the various functions of the network.

The source coding of the video and audio signals in accordance with MPEG corresponds to the fifth layer, the session layer. The definition of the individual bits in terms of the MPEG video or audio syntax can be seen as an analogy to the language alphabet. The multiplex, i.e. the preparation of the data for transport, corresponds to the manifestation of a written message put on paper and represents the transport layer. The two uppermost layers represent the generating of the programme content by means of camera and microphone or, at the decoder, by playing and viewing the video; hence they do not form part of the MPEG standards. Neither do the three lowest layers, the network layers that provide the transport path for the data, lie within the domain of MPEG. The systems required for these layers were developed by the DVB Project. The service information [ETS 468] required for the completion of the MPEG-2 system documents represents the connection between DVB

ISO/OSI	MPEG-2	
5 session layer	**compression layer**	– video and audio encoding/decoding
4 transport layer	**system layer**	**PES packet layer** – synchronisation of bit streams (e.g. video/audio)
		transport packet layer (transport MUX) or pack layer (program MUX) – multiplexing/demultiplexing – buffer management – timing – data transmission/reception

Fig. 5.3. MPEG-2 systems in the ISO/OSI layer model

and MPEG at the level of the transport layer. Just as, in the example of the letter, the precise definition of the interfaces between the layers makes it possible to replace an individual layer or several layers by others that have the same function (in the case of a postal strike, for instance, the letter completed in layer 4 can also be conveyed by a private courier service), so the MPEG-2 transport multiplex can be transmitted via any network capable of a transparent data transmission. That is exactly what was aimed for by specifying the ISO/OSI layer model.

Both layers specified in MPEG-2 are considered more closely in figure 5.3. Whilst layer 5, referred to in the ISO/OSI layer model as the "session layer" and in MPEG-2 usage as the "compression layer", includes the audio and video coding and decoding as described in chapters 3 and 4, the transport layer (or system layer) is divided into two sublayers which perform the various tasks of the multiplex. The PES packet layer is essentially responsible for the synchronisation of the elementary streams. Multiplexing/demultiplexing, buffer management, timing, and the actual transmitting and receiving of the data (which means, the transfer to or from the next-lower layer) take place in the transport packet layer (or, in the case of the program multiplex, in the pack layer).

5.3 End-to-end Synchronisation

A transmission path from the source (i.e. the camera) to the sink (i.e. the display) is shown in figure 5.4. Along the whole path, a constant transmission delay must be ensured so that the individual images of a sequence of moving

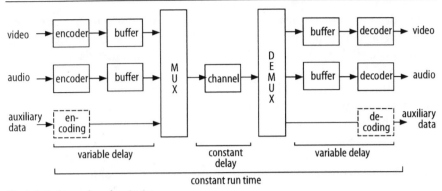

Fig. 5.4. End-to-end synchronisation

pictures can be reproduced correctly at the given frame frequency. In coding and decoding, however, delay times vary; moreover, the image delay is not, as a rule, identical with that of the corresponding audio- and auxiliary information. In order to nevertheless ensure the regular display of the images, all data must be synchronised. This task is fulfilled by the system layer.

A basic prerequisite for decoding is the recovery of the encoding clock in the decoder. An example of the method used for synchronising the decoder is shown by the block diagram in figure 5.5. Following the channel demodulation and the evaluation of the error protection, a transport stream is supplied

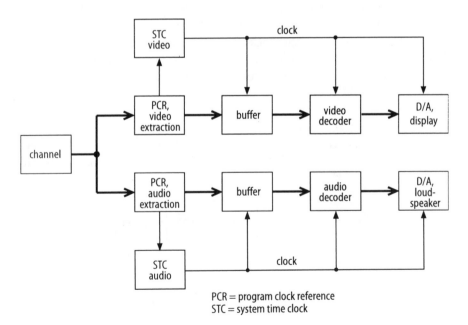

PCR = program clock reference
STC = system time clock

Fig. 5.5. Example of decoder synchronisation

to the decoder and is then fed in its entirety into the separate decoding branches for video and audio. In the case of the video branch, the first step is to eliminate all those data which are not required (e.g. audio data) and, with the aid of the "packet identification" (Packet ID, PID) of the individual packets, to extract only those packets which form part of the image information of the chosen programme. Included in these is the "program clock reference" (PCR). The "system time clock" (STC) compares itself to the reference supplied in the bit stream at least every 0.1 seconds and is corrected if necessary. In this way even the smallest deviations of the crystal-controlled 27-MHz decoder clock are compensated and the required level of precision is achieved.

The structure of the circuit in figure 5.5 with separate clocks for video and audio is an example only. It is, of course, also possible to have only one single extraction block, which generates a common clock and supplies video and audio to two separate outputs. All components belonging to the same programme have the same time reference anyway. Several different time references in one receiver are only necessary when several different programmes have to be decoded simultaneously, for example when there is a "picture in picture" feature or when a video recorder has to be fed separately.

The stabilisation of the decoder clock by the timing reference included in the bit stream is shown in figure 5.6. The "program clock reference" (PCR) is compared with the "system time clock" (STC) in the decoder which was generated by the crystal-controlled 27-MHz system clock. If variations occur, the "numerically controlled oscillator" that generates the system clock is caused to increase or decrease the required clock frequency accordingly. Apart from that, the PCR is loaded into the clock as a supporting value. Between the reference moments, defined by the MPEG-2 standard as having a maximum interval of 0.1 seconds, the clock runs free, a fact which makes a high demand on the precision of the oscillator.

The clock recovery in the decoder is a task to be performed in common for all elementary streams of a programme and is therefore carried out at the transport packet layer, the parent multiplex stage. That is why the PCR from the encoder is not yet inserted at the level of the PES bit streams, but only at

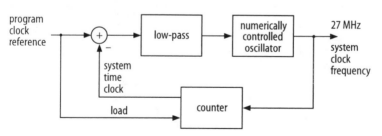

Fig. 5.6. Clock recovery in the decoder

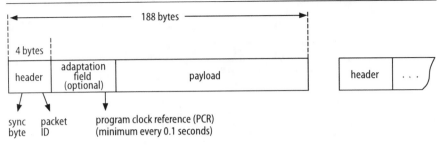

Fig. 5.7. Syntax of the transport stream (TS)

the level of the transport multiplex. A transport packet is shown in figure 5.7. Like every packet, it contains a header whose main task consists in signalling the beginning of the packet through the specified sync byte and in informing the decoder, by means of the packet ID, which type of payload the packet contains. The adaptation field may follow as an optional additional header which, among other things, contains the time reference. The header and the adaptation field contain a great amount of further important signalling information, which cannot be dealt with in detail here. The multiplexer provides for the repetition of the PCR at the prescribed intervals, depending on the data rates for the whole multiplex and the individual data streams. It can be seen that any change in the configuration of the multiplex ("remultiplexing"), for instance in a cable head end, makes a "restamping" of the PCR necessary.

Following the evaluation of the PCR, the decoder clock is in synchronisation with the encoder clock. On this basis the individual processes in the decoder, up to display and loudspeaker, can now be synchronised. This is a problem which has to be solved for each elementary stream individually. Therefore the respective "decoding time stamps" (DTSs) and "presentation time stamps" (PTSs) are to be found at the level of the PESs. Figure 5.8 shows that

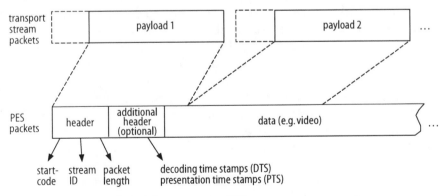

Fig. 5.8. Syntax of the packetised elementary stream (PES)

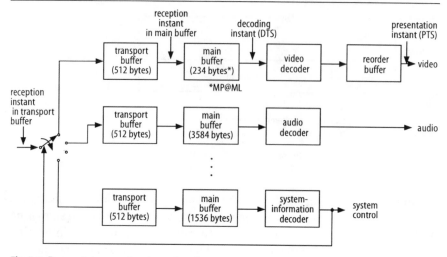

Fig. 5.9. Transport stream system target decoder

the signalling at the PES level is comparable to that at the TS level. The only addition is the indication of the packet length in the header since this length has not been fixed and can vary from packet to packet. Figure 5.9 illustrates the role of DTS and PTS. The PTS indicates at which moment, measured by the 27-MHz clock, the corresponding image should be displayed or the respective audio data be rendered audible to ensure lip sync. The DTS, on the other hand, controls the moment at which the data received must enter the decoder. This is important for the management of the data within the decoding branch.

Apart from the paths for decoding the video and audio information, figure 5.9 shows one more branch, which fulfils the functions that are vital for the system control. For example, the demultiplexing of transport packets, which is indicated by a multiple switch, is controlled by this branch, since the allocation of the packet ID is not predetermined but can be varied by the multiplexer within certain boundaries. In order to make a decoding at all possible under these conditions, the PID 0 was allocated to the so-called "program association table" (PAT) (figure 5.10). This table contains a breakdown of the multiplex into individual programmes and refers to the associated packet ID. When an MPEG-2 television receiver is switched on, only the packets with the PID 0 are evaluated to start with, and a list of the available programmes is thus compiled. It is only thereafter that the components of the chosen programme can be fed into the individual decoding branches, that the clock can be regenerated, the decoding carried out, and the video and audio information reproduced.

The MPEG-2 transport multiplex contains some further tables, which will be discussed in the next section.

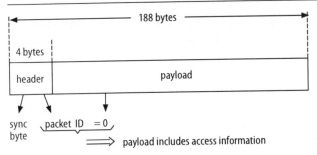

Fig. 5.10. Transmission of program association table (PAT)

5.4 Service Information

A total of four tables, containing the so-called "program-specific information" (PSI), is defined by MPEG-2 in its Systems part. Apart from this, the Systems part leaves room for the definition of further tables for particular applications. The DVB Project has taken advantage of this possibility in order to provide an automatic tuning of receivers and a support for a convenient user interface in view of the multiplicity of programmes. These additional five tables come under the heading of "service information" (SI) [ETS 468]. Both these expressions (PSI and SI) are also generally used for all nine tables. The tables that were already defined by MPEG-2 are listed as follows:

Program Association Table (PAT) (packet ID = 0)
- contains a list of all programmes in the transport multiplex and points to the packet IDs of the respective "program map tables" (PMTs)

Program Map Table (PMT)
- points to the individual packet IDs of the respective programme and in particular to the packets with the PCR
- contains the name of the programme
- contains copyright information
- etc.

Conditional Access Table (CAT) (packet ID = 1)
- contains private data for conditional access

Network Information Table (NIT)
- contains private data, i.e. orbit position, transponder number, ...

The contents of the last two tables were not specified by MPEG-2 but by the DVB Project. All these tables contain information that is necessary for the technical realisation of the data transmission. In this way the PMT complements the PAT by providing more detailed information about the individual programmes required for the decoding. The NIT contains the information which is required for the automatic tuning of the receivers to the corresponding programme. Finally, the CA table supplies the information required for the scrambling and descrambling of programmes or programme parts (see chapter 8).

The tables defined by the DVB Project in addition to the above ones supply information intended for the viewers, for example, information facilitating an electronic programme guide or information regarding the control of video recorders [ETR 211].

Bouquet Association Table (BAT)
- contains information about the bouquet of programmes of one provider, even when these are distributed via various transmission paths

Service Description Table (SDT)
- describes the programmes offered
- contains, for example, details about the content provider

Event Information Table (EIT)
- contains programme tables similar to those in the traditional programme guides
- contains an identification for each type of programme
- contains a classification of the programmes with regard to the suitability for a particular age group of viewers

Time and Date Table (TDT)
- contains the actual time

Running Status Table (RST)
- shows whether a particular programme is running, not yet running or about to commence in the immediate future and, in this way, enables the control of video recorders as well as other functions.

This service information is an important element of the DVB programme service because it enables the user to find his or her way with ease through the considerably increased number of programmes that have become available in the digital world. User guidance is bound to be much more comprehensive compared to current teletext services. For example, the receiver could be programmed to the viewing preferences of the owner (variety, crime series, magazine programme, etc.) and, when switched on, could refer to that type of programme running or starting shortly. Traditional teletext will also be offered by DVB [ETS 472] in order to integrate DVB transmissions into present infrastructures. Finally, the DVB Project has developed a new technique for subtitling [ETS 743] in which the subtitles are transmitted as text and only inserted into the picture in the receiver. This enables the viewer to choose, for example, the size and type of script most suitable to his or her personal requirements.

6 Forward Error Correction (FEC) in Digital Television Transmission

In MPEG source coding (see chapters 3 and 4) redundancy and irrelevance are eliminated from the digital signal. However, this means that the signal is more vulnerable to disturbances in the transmission path. If a digital television image which has not been source-coded is transmitted at a data rate of, for example, 270 Mbit/s and one bit has been altered during transmission, only one pixel of the image will be affected in most cases. However, if an MPEG-coded image is transmitted, a bit error can result in at least one erroneous macroblock, if not in several.

A bit error can occur, for example, if because of too much noise a detection threshold in the demodulator is crossed and the value received is thus assigned to a different, incorrect symbol (see chapter 7).

Therefore channel coding is required which provides for forward error correction [FEC]. This enables the receiver to correct the errors which have occurred in the transmission path.

In chapter 6 the two relevant methods of error protection in the transmission of digital television will be introduced. For the sake of simplicity, examples will be shown which, although their parameters are not those of the European DVB standard (cf. chapters 9 to 11), will nevertheless enable the reader to easily understand the principle of error correction. It is relatively easy to transfer these parameters to those actually used in the DVB standard (see section 6.4.3).

6.1 Basic Observations

In this section there will be a short introduction to the basics of error correction, without going too deeply into the theory. A more comprehensive insight into the theoretical and mathematical aspects can be obtained from [SWEENEY].

The principle of the transmission with error-protection coding is shown in figure 6.1.

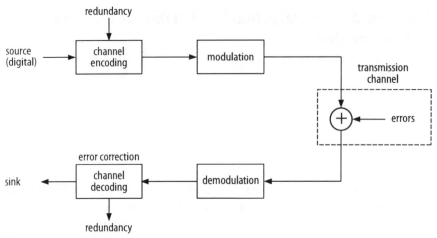

Fig. 6.1. Transmission with error-protection coding

At the transmitter, redundancy is added to the source-coded digital signal in the channel encoder. This is to enable the channel decoder in the receiver to correct any errors. The addition of the redundancy (resulting in the transmission of additional data not included in the source signal) leads to an increase in the data volume to be transmitted. This must be taken into account when choosing the type of modulation in order not to exceed the maximum possible channel capacity.

The digitally modulated signal is overlaid with errors within the transmission channel which are caused by the invalidation of one or more bits. A '1' becomes a '0' and vice versa. The task of channel coding in the receiver is to find the position of the incorrect bits by the evaluation of the redundancy, which is possibly also affected by transmission errors, and to invert these. The added redundancy is then removed.

Figure 6.2 outlines the various types of error which can occur.

Errors occurring singly within a data stream are called bit errors. An n-bit burst error is defined by a block of the length of n bits in which at least the first bit and the last bit are erroneous. Individual bits within this block, however, are not necessarily erroneous. Finally, a symbol error denotes one erroneous symbol, which for example in figure 6.2 has a length of 8 bits (1 byte). Within this symbol up to 8 bit errors can occur in a random constellation.

With the knowledge of the actual types of error occurring in the channel, various codes can be constructed which will successfully correct the most common error types, but less successfully the less common error types. Figure 6.3 gives an overview of the most common classes of codes. It is neither complete nor is every code dealt with individually here (see [CLARK]).

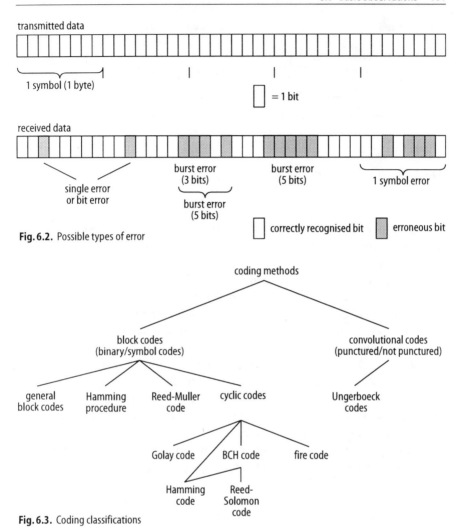

Fig. 6.2. Possible types of error

Fig. 6.3. Coding classifications

The most important criterion is the distinction between the block code and the convolutional code. In the case of block codes the input data stream is divided into blocks of the fixed length m, where m denotes the number of symbols. Such symbols can either be comprised of one bit or, as in the present case, of several bits. In the first case the codes are binary codes and in the second case, symbol codes. A symbol-oriented code is particularly well suited to the correction of symbol errors, concerning which it is of no significance which bit of a symbol is erroneous. The error correction must then not only find the erroneous symbol but also determine the original value of the symbol, whereas with binary codes only the erroneous bit needs to be inverted.

(n,m) block code

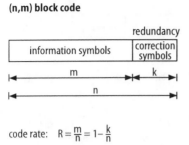

code rate: $R = \frac{m}{n} = 1 - \frac{k}{n}$

convolutional code

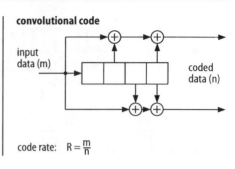

code rate: $R = \frac{m}{n}$

Fig. 6.4. Block codes and convolutional codes

In the block codes the calculated redundancy is appended to the actual m information positions in the k correction positions so that finally a block with a length of $n = m + k$ is transmitted. The code rate indicates the ratio between the information m and the transmission symbols n.

The convolutional code differs from the block code in that a binary shift register is always used. There is no division into predetermined segments of the input data stream, but the input information is spread over several output data. This takes place through storage of the input data in a shift register and generation of the output data through a combination of various taps at the shift register. It is because of this type of coding that the convolutional codes are destined for the correction of individual bit errors. The code rate is here defined as the ratio of the number of input bits to the number of output bits which are generated at the same step. As in the case of the block codes, the code rate is always <1, as otherwise no redundancy would have been added. Figure 6.4 shows a comparison of the principles of the block codes with those of the convolutional codes.

Finally, the "bit-error rate" (BER) should be introduced at this point. This is the term used to denote the proportion of erroneously decoded bits to the total number of bits received. Accordingly, the term "symbol-error rate" (SER) denotes the proportion of erroneously decoded symbols to the total number of symbols received.

6.2 Reed-Solomon Codes

Reed-Solomon codes are symbol-oriented block codes. That means, the error correction must not only recognise which symbol of the block of length n is erroneous, but also compute the value of the original symbol. As a rule, one symbol contains 8 bits. To make the following section easier to understand, examples will use 3 bits per symbol.

The Reed-Solomon encoding and decoding require a considerable amount of computation. As we are moving within a finite range of numbers which has 256 elements in the case of 8 bits per symbol or, correspondingly, 8 elements in the case of 3 bits per symbol, the arithmetic to be employed must guarantee that the result of an operation using elements from this number space will again be an element from the same number space. A prerequisite is that all arithmetical operations, i.e. addition, multiplication, subtraction, division, etc. must be valid. It is the arithmetic of the Galois field which is used for the Reed-Solomon code. Therefore it is necessary to understand the system of the Galois field in order to understand the way the Reed-Solomon code works. That is why the following subsection will be an introduction to the arithmetic of the Galois field, in which great care has been taken to choose examples that are easy to understand. Those readers who require more detail and who are interested in the mathematical proofs of the matters to be discussed should consult [SWEENEY], [CLARK]. The examples of the Reed-Solomon code shown here are based on an encoding and a decoding within the frequency domain. This procedure, although corresponding exactly to the definition of the Reed-Solomon code, is not actually used in practice. However, it is considerably easier to understand than the respective procedure with encoding and decoding in the time domain, so that it is more suitable for an introduction to the functional principal of the Reed-Solomon code. The transfer to the procedure which works in the time domain is relatively easy (see section 6.2.5 or [CLARK]). But for this, too, the most important precondition is to understand the Galois field.

6.2.1 Introduction to the Arithmetic of the Galois Field

The Galois field is defined as follows:

A Galois field, short form GF(q) (here always $q = 2^w$, $w = 3$), is a finite field which contains a set of q different elements. The following rules govern the elements of this field:

- *The arithmetical operations of addition and multiplication are so defined that the combination of two field elements will always result in a field element.*
- *The neutral element of addition (0) as well as the neutral element of multiplication (1) are contained within the field.*
- *For each element β an additive inverse element (−β) and a multiplicative inverse element $β^{-1}$ exist, so that $β + (− β) = 0$ and $β · β^{-1} = 1$. The subtraction and division of field elements are defined in this way.*
- *The associative and commutative laws apply.*

At this point a short definition of the modulo operation is appropriate as this is not known to all readers. The modulo operation of two numbers results in the residual quantity of the whole-number division of the two numbers. For example, if the number 7 is divided by 3, the result is 2 and the residual quantity is 1: $7 = 2 \cdot 3 + 1$, or 7 mod 3 = 1.

The result of a modulo operation is always smaller than the argument of the modulo operation, for example a modulo-2 operation can only have 1 or 0 as a result.

In the same way as the modulo operation can be carried out with whole numbers it can be applied to polynomials. The result of a modulo operation with two polynomials is the residual quantity (a polynomial!) which results from a polynomial division. The degree of this (residual) polynomial is therefore, by definition, smaller than the degree of the polynomial through which it was divided.

On the basis of the Galois field as defined above and with the knowledge of the modulo operation we can now assign the following properties to the $GF(2^w)$:

(1) The elements of the $GF(2^w)$ are polynomials of degree <w.
(2) The coefficients of the polynomials are 0 or 1.[1]
(3) The addition of two elements is the modulo-2 addition, that is, the EXCLUSIVE-OR linkage (XOR) of the polynomial coefficients which correspond to each other. Therefore an element is equal to its inverse element and the following is valid: $\beta + (-\beta) = \beta + \beta = 0$.
(4) The multiplication of two elements is the multiplication of the polynomials (taking property 3 into account) and the subsequent modulo operation with the generator polynomial $g(x)$ of the Galois field $GF(2^w)$.
(5) The generator polynomial $g(x)$ of the Galois field $GF(2^w)$, comprised of the coefficients 0 or 1, can be freely selected; however, it must be of degree w and irreducible, i.e. incapable of being factorised.[2]

These properties can be illustrated by the following example:

Let the two polynomials

$$\beta_1 = x^2 + x + 1 \quad \in GF(2^3), \quad \text{degree } (\beta_1) = 2 < w = 3 \tag{6.1a}$$

$$\beta_2 = x^2 + 1 \quad \quad \in GF(2^3), \quad \text{degree } (\beta_2) = 2 < w = 3 \tag{6.1b}$$

be defined. Their degree is 2 and their coefficients are 0 or 1. Herewith properties 1 and 2 are satisfied. The addition is now the modulo-2 addition of the corresponding coefficients. Therefore the following is valid:

---

[1]   The coefficients are elements of the $GF(2)$.
[2]   Even the generator polynomial has coefficients from the $GF(2)$. 'Irreducible' means that the polynomial cannot be broken down into a product of further polynomials unless these have coefficients which are not derived from the $GF(2)$.

$$\beta_1 + \beta_2 = (x^2 + x + 1) + (x^2 + 1)$$
$$= x^2 + x^2 + x + 1 + 1 \tag{6.2}$$
$$= x$$

The coefficient of x^2 is equal to 1 in both β_1 and β_2. The result of the modulo-2 addition (or XOR operation) is therefore 0. The same is the case for x^0. For the multiplication, a generator polynomial must be defined which has the properties as defined under (5). For this there are tables, for example in [FURRER]. The following polynomial was selected:

$$g(x) = x^3 + x + 1 \quad \text{irreducible, degree } [g(x)] = w = 3 \tag{6.3}$$

The polynomials β_1 and β_2 are now multiplied with each other while still satisfying property 3. The result is then subjected to the modulo operation with the generator polynomial $g(x)$ (once again taking property 3 into account). The result of this modulo-$g(x)$ operation represents the multipliers of β_1 and β_2 in the Galois field.

$$\beta_1 \cdot \beta_2 = [(x^2 + x + 1)(x^2 + 1)] \bmod (x^3 + x + 1)$$
$$= (x^4 + x^3 + x^2 + x^2 + x + 1) \bmod (x^3 + x + 1) \tag{6.4}$$
$$= (x^4 + x^3 + x + 1) \bmod (x^3 + x + 1)$$
$$= x^2 + x$$

As expected, the result is yet again an element of $GF(2^w)$. It has the properties 1 and 2. The modulo operation with the generator polynomial of degree w ensures that the degree of the resulting polynomial is $<w$. One of the properties of all Galois fields is that for each field a so-called primitive field element α exists, the exponents of which can be used to generate or represent all other elements of the field. The Galois field can be completely and uniquely defined by this primitive field element and the generator polynomial as defined by property 5. In most cases the primitive field element α corresponds to the element of the Galois field which is described by the polynomial $\alpha = x$.

On the assumption that $\alpha = x$ is the primitive field element and $g(x) = x^3 + x + 1$ is the generator polynomial, we are listing below – by way of an example – the elements of the respective Galois field $GF(2^3)$.

The calculation of the field elements is as follows: the first element is zero, which in accordance with the definition must be included in the Galois field. All further elements can be represented by the involution of α, for example, α^0, α^1 and α^2. When commencing with the exponent α^3 one has to bear in mind that the degree of the resulting polynomial must be <3. Here the modulo operation with $g(x)$ must be used and this will lead to the result $\alpha^3 = x + 1$.

Table 6.1. Calculation and binary representation
of the elements in a Galois field

Calculation	Binary representation
0	000
$\alpha^0 = 1$	001
$\alpha^1 = x$	010
$\alpha^2 = x^2$	100
$\alpha^3 = x^3 = x+1$	011
$\alpha^4 = x^2+x$	110
$\alpha^5 = x^3+x^2 = x^2+x+1$	111

The calculation of α^4, α^5 and α^6 is done in the same way. When calculating α^7 it can be seen that $\alpha^7 = \alpha^0$, which means that the Galois field is cyclic. $\alpha^x = \alpha^{x \bmod 7}$ or, more generally, $\alpha^x = \alpha^{x \bmod (2^w-1)}$. The binary notation of the elements of the Galois field can be found in the second column of table 6.1. One element of the Galois field is represented here by 3 bits, one bit each for one coefficient of the polynomial corresponding to the field element. The element which was generated by α^6 is taken as an example. After computing α^6 (by applying the modulo-$g(x)$ operation) the result is $\alpha^6 = x^2 + 1$. The coefficient of x^2 is 1, that of x^1 is 0, and that of x^0 is 1. Therefore the binary notation of the field element is '101'.[3]

On the basis of table 6.1 it is easy to perform the operations of addition and multiplication:

- The addition of two elements corresponds to the EXCLUSIVE-OR linkage (XOR) of the elements in the binary notation.
- The multiplication is the addition and the subsequent modulo-7 operation of the exponents of the powers of α.[4]
- To convert the binary mode to the exponential mode and vice versa, it is necessary to use a table like the one reproduced above.

It is now necessary to reiterate why all calculations are made in the Galois field rather than in a real or natural space of numbers: the aim is to define an error-correcting code which operates on the basis of symbols. Each symbol can take on a certain number of 2^w values. Each calculation or arithmetical link of two such symbols must guarantee that the result of this operation lies within the original range of values.

Provided that the above requirement is taken into account, a discrete Fourier transform (DFT) can now be defined in the Galois field. The mapping of a set into another set, for example the transfer from the time domain to the frequency domain, must now also be effected in a way which ensures that again

---

[3] Also other binary representations are possible. For this see [SWEENEY].
[4] In general a modulo-(2^w-1) operation should be carried out in $GF(2^w)$.

only the elements of the Galois field are present in the image domain. The corresponding formulae for the DFT and its inverse function (IDFT) will only be mentioned here, without giving the mathematical derivations. These can be found, for example, in [SWEENEY]. Here again the Galois field is assumed as being $GF(2^w)$, and its primitive element as being α.

DFT in the Galois field $GF(2^w)$:

$$A_l = \sum_{i=0}^{N-1} a_i \alpha^{+il} \qquad N = 2^w - 1 \quad l = 0 \dots N-1 \tag{6.5}$$

IDFT in the Galois field $GF(2^w)$:

$$a_l = \sum_{i=0}^{N-1} A_i \alpha^{-il} \qquad N = 2^w - 1 \quad l = 0 \dots N-1 \tag{6.6}$$

When, in the following, the expressions 'time domain' and 'frequency domain' are used, it is in order to differentiate between the two domains of transformation and inverse transformation.

The definition of the Reed-Solomon code is derived from another characteristic of the Galois field which is explained by the following 'theorem of roots and spectral components' and will be illustrated by an example in which the previously defined DFT is used.

The polynomial

$$a(X) = a_{l-1}X^{l-1} + \dots + a_i X^i + \dots + a_2 X^2 + a_1 X + a_0 \tag{6.7}$$

from $GF(2^w)$ has a root α^i if and only if the spectral component A_i, equals zero.

A prerequisite is that the coefficients $a_0 \dots a_{l-1}$ are elements of the Galois field $GF(2^w)$ and that X is the argument of the polynomial $a(X)$, i.e. for X, too, an element of the Galois field $GF(2^w)$ must be substituted. So if α^i is substituted for X and if $a(X) = a(\alpha^i) = 0$ results, it follows that the spectral component A_i, i.e. the coefficient A_i of the transformed polynomial, equals zero.

Strictly speaking, this theorem again describes (6.5), in which a sum is formed for a fixed α^l which corresponds exactly to (6.7). If that sum is zero, i.e. if α^l is a root, then and only then is A_l also equal to zero.

For the inverse transformation the following is valid:

The transformed polynomial

$$A(Z) = A_{l-1}Z^{l-1} + \dots + A_i Z^i + \dots + A_2 Z^2 + A_1 Z + A_0 \tag{6.8}$$

has a root α^{-i} if and only if the component a_i of the inverse transform, i.e. the coefficient a_i of the polynomial $a(X)$, equals zero.

Here the coefficients $A_0 \ldots A_{l-1}$ are again elements of the Galois field $GF(2^w)$ and Z is the argument of the polynomial $A(Z)$. If $Z = \alpha^{-i}$ is substituted and results in $a(Z) = A(\alpha^{-i}) = 0$, the coefficient a_i of the inverse transformed polynomial $a(X)$ equals zero.

The specimen calculation below will again be in $GF(2^w)$. For convenience, let us define the polynomial

$$a(X) = \alpha^2 X^6 + \alpha^3 X^5 + \alpha^6 X^4 + \alpha^5 X^2 + \alpha^4 . \tag{6.9}$$

It can be seen that $a_3 = a_1 = 0$. In accordance with the above theorem the transformed polynomial $A(Z)$ must show zero roots for $Z = \alpha^{-3}$ and $Z = \alpha^{-1}$. In order to check this, first the spectral components $A_0 \ldots A_6$ are computed in accordance with (6.5). For the addition and transformation of the two representations of the elements table 6.1 can again be consulted.

$$
\begin{aligned}
A_0 &= \alpha^4 + 0 + \alpha^5 + 0 + \alpha^6 + \alpha^3 + \alpha^2 \\
&= \alpha^4 + \alpha^5 + \alpha^6 + \alpha^3 + \alpha^2 = \alpha^3 \\
A_1 &= \alpha^4 + 0\alpha + \alpha^5\alpha^2 + 0\alpha^3 + \alpha^6\alpha^4 + \alpha^3\alpha^5 + \alpha^2\alpha^6 \\
&= \alpha^4 + \alpha^0 + \alpha^3 + \alpha^1 + \alpha^1 = \alpha^2 \\
A_2 &= \alpha^4 + 0\alpha^2 + \alpha^5\alpha^4 + 0\alpha^6 + \alpha^6\alpha^8 + \alpha^3\alpha^{10} + \alpha^2\alpha^{12} \\
&= \alpha^4 + \alpha^2 + \alpha^0 + \alpha^6 + \alpha^0 = \alpha^5 \\
A_3 &= \alpha^4 + 0\alpha^3 + \alpha^5\alpha^6 + 0\alpha^9 + \alpha^6\alpha^{12} + \alpha^3\alpha^{15} + \alpha^2\alpha^{18} \\
&= \alpha^4 + \alpha^4 + \alpha^4 + \alpha^4 + \alpha^6 = \alpha^6 \\
A_4 &= \alpha^4 + 0\alpha^4 + \alpha^5\alpha^8 + 0\alpha^{12} + \alpha^6\alpha^{16} + \alpha^3\alpha^{20} + \alpha^2\alpha^{24} \\
&= \alpha^4 + \alpha^6 + \alpha^1 + \alpha^2 + \alpha^5 = \alpha^1 \\
A_5 &= \alpha^4 + 0\alpha^5 + \alpha^5\alpha^{10} + 0\alpha^{15} + \alpha^6\alpha^{20} + \alpha^3\alpha^{25} + \alpha^2\alpha^{30} \\
&= \alpha^4 + \alpha^1 + \alpha^5 + \alpha^0 + \alpha^4 = \alpha^2 \\
A_6 &= \alpha^4 + 0\alpha^6 + \alpha^5\alpha^{12} + 0\alpha^{18} + \alpha^6\alpha^{24} + \alpha^3\alpha^{30} + \alpha^2\alpha^{36} \\
&= \alpha^4 + \alpha^3 + \alpha^2 + \alpha^5 + \alpha^3 = \alpha^6
\end{aligned}
\tag{6.10}
$$

The transformed polynomial therefore gives

$$A(Z) = \alpha^6 Z^6 + \alpha^2 Z^5 + \alpha Z^4 + \alpha^6 Z^3 + \alpha^5 Z^2 + \alpha^2 Z + \alpha^3 . \tag{6.11}$$

The substitution of the zero roots to be checked results in

$$A(\alpha^{-1}) = \alpha^{6}\alpha^{-6} + \alpha^{2}\alpha^{-5} + \alpha\alpha^{-4} + \alpha^{6}\alpha^{-3} + \alpha^{5}\alpha^{-2} + \alpha^{2}\alpha^{-1} + \alpha^{3}$$

$$= \alpha^{0} + \alpha^{-3} + \alpha^{-3} + \alpha^{3} + \alpha^{3} + \alpha^{1} + \alpha^{3}$$

$$= \alpha^{0} + \alpha^{1} + \alpha^{3} = 0 \tag{6.12}$$

$$A(\alpha^{-3}) = \alpha^{6}\alpha^{-18} + \alpha^{2}\alpha^{-15} + \alpha\alpha^{-12} + \alpha^{6}\alpha^{-9} + \alpha^{5}\alpha^{-6} + \alpha^{2}\alpha^{-3} + \alpha^{3}$$

$$= \alpha^{-5} + \alpha^{-6} + \alpha^{-4} + \alpha^{-3} + \alpha^{-1} + \alpha^{-1} + \alpha^{3}$$

$$= \alpha^{2} + \alpha^{1} + \alpha^{3} + \alpha^{4} + \alpha^{3} = 0$$

QED.

6.2.2 Definition of the RS Code and the Encoding/Decoding in the Frequency Domain

A t-error-correcting RS code, with $q = 2^{w}$ symbol values and a block length $n = q - 1$, is the set of all words whose spectrum within $GF(q)$ is zero in $k = 2t$ consecutive components.

An RS code vector can now be generated using the 'theorem of roots and spectral components' as given in the previous section.

- First of all, a particular Galois field $GF(2^{w})$ is chosen, which is defined by its generator polynomial $g(x)$ and its primitive field element α.
- With this, the block length n ($n = q - 1$) is determined.
- After choosing the error-correction possibility t of the code, k – and therefore also m – is known (remember: $n = m + k$).
- Starting from the spectrum, the k positions $C_0 \ldots C_{2t-1}$ of the block with length n are set to zero. Therefore there are $2t$ consecutive zeros.
- The information symbols are placed in the m positions $C_{2t} \ldots C_{n-1}$.
- If, finally, an IDFT is performed in accordance with (6.6), this operation results in a valid transmittable RS code vector $c(X)$ which corresponds to the above definition.

This procedure is shown again clearly in figure 6.5.

In the transmitting channel the (polynomial) code vector $c(X)$ transmitted is overlaid with a (polynomial) error vector $e(X)$ which, however, is unknown to the receiver.

At the receiving end the signal spectrum $R(Z)$, which may have received an erroneous vector $r(X)$, can be computed using a DFT. At this point it is already possible to decide whether an error occurred during the transmission. If the positions $E_0 \ldots E_{2t-1}$ are zero, a valid RS code vector has been received (see the above definition of the RS code) and the positions $R_{n-1} \ldots R_{2t}$ contain the original information symbols. Hence, the transmission was error-free. If the positions $E_0 \ldots E_{2t-1}$, or even only one of them, are not zero, the spectrum

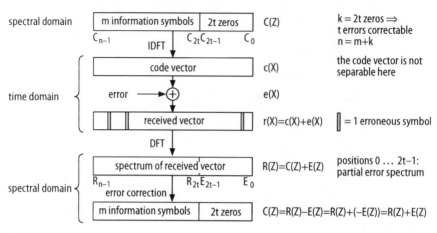

Fig. 6.5. Encoding and decoding of an RS code vector in the frequency domain

$R(Z)$ of the received vector has been additively overlaid with the spectrum $E(Z)$ of the error vector. It is the task of the subsequent error correction to define the spectrum of the error vector, so that

$$R(Z) = C(Z) + E(Z) \qquad (6.13)$$

again results in

$$C(Z) = R(Z) - E(Z) = R(Z) + [-E(Z)] = R(Z) + E(Z) \qquad (6.14)$$

if $E(Z)$ is known.[5] To understand this clearly we must remember that the positions $R_0 \ldots R_{2t-1}$ correspond to $E_0 \ldots E_{2t-1}$, as here there were originally zeros. Since the spectrum of the error-vector polynomial additively overlays the zeros, a number of the spectral coefficients of the error vector (namely $E_0 \ldots E_{2t-1}$) are available in the receiver immediately after the DFT, ready to be used by the subsequent error correction.

6.2.3 Error Correction Using the RS Code

The task of the error correction is to compute all positions of $E(Z)$ from the known positions $E_0 \ldots E_{2t-1}$ of the spectrum $E(Z)$ of the error vector $e(X)$, so that the only thing left to do is to additively combine $E(Z)$ with $R(Z)$ in order to receive the original information $C(Z)$ (see (6.14)). To solve this problem an error-locator polynomial $\lambda(X)$ is introduced to begin with. This is so defined that at the position at which the error-vector polynomial $e(X)$ has a coeffi-

---

[5] In this case the addition of two polynomials corresponds of course again to the modulo-2 linkage of the coefficients which correspond to each other.

cient not equal to zero, i.e. at the position at which an error has occurred, the corresponding coefficient λ_i of $\lambda(X)$ is zero. The precise knowledge of this position, however, is not required for the time being. The transformed error-locator polynomial $\lambda(X)$ is denoted as $\Lambda(Z)$. Therefore the product of λ_i and e_i, as well as its DFT, is:

$$\lambda_i e_i = 0 \quad \xrightarrow{\ DFT\ } \quad \sum_{i=0}^{n-1} \Lambda_i E_{l-1} = 0 \qquad (6.15)^6$$

The following simplifications can be used in general without restriction:

- $\Lambda_0 = 1$
- $\Lambda_i = 0$ for $i > t$: a maximum of t errors should be able to be corrected, therefore there are a maximum of t positions at which $\lambda_i = 0$, thus $\Lambda(Z)$ is the maximum degree of t.[7]

$\Rightarrow i = 0 \ldots t$

$\Rightarrow l = t \ldots 2t - 1$, when at first only $E_0 \ldots E_{2t-1}$ are known.

Using these simplifications, (6.15) can be rewritten as the following set of equations:

$$
\begin{aligned}
\Lambda_0 E_t \quad &+ \Lambda_1 E_{t-1} \quad + \ldots + \quad \Lambda_t E_0 \quad = 0 \\
\Lambda_0 E_{t+1} \quad &+ \Lambda_1 E_t \quad + \ldots + \quad \Lambda_t E_1 \quad = 0 \\
&\qquad\qquad\qquad\qquad\vdots \\
\Lambda_0 E_{2t-1} \quad &+ \Lambda_1 E_{2t-2} \quad + \ldots + \quad \Lambda_t E_{t-1} \quad = 0
\end{aligned}
\qquad (6.16)
$$

In this so-called key equation for $\Lambda(Z)$ both $E_0 \ldots E_{2t-1}$ and Λ_0 are known. It comprises t equations with t unknowns and can therefore be solved.

In the following the above simplifications shall apply. Moreover it can be assumed that $\Lambda_0 \ldots \Lambda_t$ are known after the solution of the set of equations. Then (6.15) can be once again rewritten as

$$\sum_{i=0}^{n-1} \Lambda_i E_{l-1} = \Lambda_0 E_1 + \sum_{i=1}^{n-1} \Lambda_i E_{l-1} = 0 \quad \Rightarrow \quad E_1 = \sum_{i=1}^{t} \Lambda_i E_{l-1}. \qquad (6.17)^8$$

On the basis of this equation each E_l can be recursively computed from $E_{l-1} \ldots E_{l-t}$. This directly leads to a feedback shift register implementation (recursive extension) in accordance with figure 6.6.

First, the registers are loaded with $E_t \ldots E_{2t-1}$ in accordance with $R_t \ldots R_{2t-1}$. With every cycle, the following E_l is now to be found at the output. All coeffi-

---

[6] In the space of real numbers the summation shown here is also known as discrete convolution.

[7] See 'theorem of roots and spectral components'.

[8] Taking into account $\Lambda_0 = 1$, $-E_l = E_l$ and $i = 1 \ldots t$.

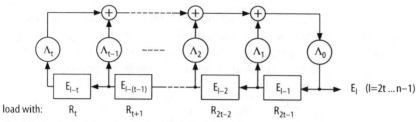

load with:

Fig. 6.6. Feedback shift register for the calculation of $E_{2t} \ldots E_{n-1}$

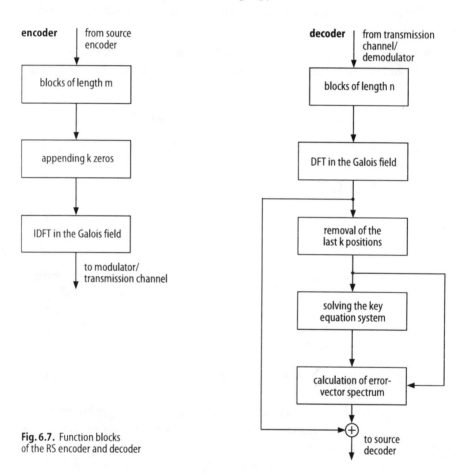

Fig. 6.7. Function blocks
of the RS encoder and decoder

cients of $E(Z)$ can be calculated in this way. By adding $R(Z)$ and $E(Z)$ one finally arrives at $C(Z)$ (see figure 6.5).

Figure 6.7 summarises the procedure of encoding and decoding Reed-Solomon code vectors in the frequency domain.

6.2.4 Examples of Encoding/Decoding in the Frequency Domain

We shall now explain, by using a simple example, how the Reed-Solomon code works. For a better understanding of the process the reader is referred to figure 6.5. We are dealing with a $(7,3)$-RS code which is defined in $GF(2^3)$. $\alpha = x$ is defined as the primitive field element, the generator polynomial of the Galois field is $g(x) = x^3 + x + 1$. The code can correct two errors and thus transmit 3 information symbols. The three information symbols α^5, α^2 and α^3 which are to be transmitted are written in the positions $C_6 \ldots C_4$. The last four positions $C_3 \ldots C_0$ are set to zero.

$$\left.\begin{array}{l} C_0 \ldots C_3 = 0 \\ C_4 = \alpha^3 \\ C_5 = \alpha^2 \\ C_6 = \alpha^5 \end{array}\right\} \Rightarrow C(Z) = \alpha^5 Z^6 + \alpha^2 Z^5 + \alpha^3 Z^4 \tag{6.18}$$

By an IDFT in accordance with (6.6) the transmitting code vector $c(X)$ is obtained.

$$\left.\begin{array}{l} c_0 = \alpha^3 + \alpha^2 + \alpha^5 = 0 \\ c_1 = \alpha^3 \alpha^{-4} + \alpha^2 \alpha^{-5} + \alpha^5 \alpha^{-6} = \alpha^4 \\ c_2 = \alpha^3 \alpha^{-8} + \alpha^2 \alpha^{-10} + \alpha^5 \alpha^{-12} = 0 \\ c_3 = \alpha^3 \alpha^{-12} + \alpha^2 \alpha^{-15} + \alpha^5 \alpha^{-18} = \alpha^5 \\ c_4 = \alpha^3 \alpha^{-16} + \alpha^2 \alpha^{-20} + \alpha^5 \alpha^{-24} = \alpha^6 \\ c_5 = \alpha^3 \alpha^{-20} + \alpha^2 \alpha^{-25} + \alpha^5 \alpha^{-30} = \alpha \\ c_6 = \alpha^3 \alpha^{-24} + \alpha^2 \alpha^{-30} + \alpha^5 \alpha^{-36} = \alpha^4 \end{array}\right\} \Rightarrow c(X)$$

$$c(X) = \alpha^4 X^6 + \alpha X^5 + \alpha^6 X^4 + \alpha^5 X^3 + \alpha^4 X \tag{6.19}$$

In the channel, the error $e(X)$ then overlays the code vector, and the result is the received vector $r(X)$:

$$e(X) = \alpha^2 X^5 + \alpha X \tag{6.20}[9]$$
$$\Rightarrow r(X) = c(X) + e(X) = \alpha^4 X^6 + \alpha^4 X^5 + \alpha^6 X^4 + \alpha^5 X^3 + \alpha^2 X.$$

The spectrum $R(Z)$ of the received vector is calculated in accordance with (6.5) as

---

[9] $e(X)$ has here been chosen at random.

$$R_0 = \alpha^2 + \alpha^5 + \alpha^6 + \alpha^4 + \alpha^4 = \alpha^4$$

$$R_1 = \alpha^2\alpha + \alpha^5\alpha^3 + \alpha^6\alpha^4 + \alpha^4\alpha^5 + \alpha^4\alpha^6 = \alpha^6$$

$$R_2 = \alpha^2\alpha^2 + \alpha^5\alpha^6 + \alpha^6\alpha^8 + \alpha^4\alpha^{10} + \alpha^4\alpha^{12} = \alpha^2$$

$$R_3 = \alpha^2\alpha^3 + \alpha^5\alpha^9 + \alpha^6\alpha^{12} + \alpha^4\alpha^{15} + \alpha^4\alpha^{18} = \alpha^6$$

$$R_4 = \alpha^2\alpha^4 + \alpha^5\alpha^{12} + \alpha^6\alpha^{16} + \alpha^4\alpha^{20} + \alpha^4\alpha^{24} = \alpha^4$$

$$R_5 = \alpha^2\alpha^5 + \alpha^5\alpha^{15} + \alpha^6\alpha^{20} + \alpha^4\alpha^{25} + \alpha^4\alpha^{30} = \alpha^2$$

$$R_6 = \alpha^2\alpha^6 + \alpha^5\alpha^{18} + \alpha^6\alpha^{24} + \alpha^4\alpha^{30} + \alpha^4\alpha^{36} = 0$$

$$\left. \right\} \Rightarrow R(Z)$$

$$R(Z) = \alpha^2 Z^5 + \alpha^4 Z^4 + \alpha^6 Z^3 + \alpha^2 Z^2 + \alpha^6 Z + \alpha^4 \tag{6.21}$$

As the last four positions $R_0 \ldots R_3$ of the spectrum $R(Z)$ are not all zero, they are bound to be overlaid with an error vector whose positions $E_0 \ldots E_3$ directly correspond to the received and transformed positions $R_0 \ldots R_3$. Using $E_0 \ldots E_3$ we can state and solve the key equation.

$$\left. \begin{array}{l} \Lambda_0\alpha^2 + \Lambda_1\alpha^6 + \Lambda_2\alpha^4 = 0 \\ \Lambda_0\alpha^6 + \Lambda_1\alpha^2 + \Lambda_2\alpha^6 = 0 \end{array} \right\} \Rightarrow \Lambda_0 = 1, \ \Lambda_1 = \alpha^6, \ \Lambda_2 = \alpha^6 \tag{6.22}$$

The values of $\Lambda_0 \ldots \Lambda_2$ determine the coefficients in the feedback branches of the shift register (see figure 6.8).

If the latter is loaded with E_3 and E_2, the spectral coefficients at the output are $E_4 \ldots E_6$ and thus part of the error vector. They result in $E_4 = \alpha_6$, $E_5 = {}_0$ and $E_6 = \alpha_5$. Finally, the original information can be reconstructed if $E(Z)$ and $R(Z)$ are known:

$$E(Z) = \alpha^5 Z^6 + 0Z^5 + \alpha^6 Z^4 + \alpha^6 Z^3 + \alpha^2 Z^2 + \alpha^6 Z + \alpha^4$$

$$R(Z) = 0Z^6 + \alpha^2 Z^5 + \alpha^4 Z^4 + \alpha^6 Z^3 + \alpha^2 Z^2 + \alpha^6 Z + \alpha^4 \tag{6.23}$$

$$C(Z) = E(Z) + R(Z) = \alpha^5 Z^6 + \alpha^2 Z^5 + \alpha^3 Z^4$$

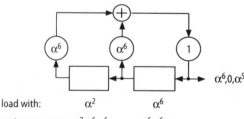

load with: α^2 α^6

register contents: $\alpha^2,\alpha^6,\alpha^6$ $\alpha^6,\alpha^6,0$

Fig. 6.8. Example of the structure of a feedback shift register for the definition of $E_4 \ldots E_6$

6.2.5 Encoding and Decoding in the Time Domain

The procedure introduced so far performs the RS encoding and decoding in the frequency domain. The advantage of this procedure is that it can be easily understood. It was chosen to demonstrate the basic modus operandi of the RS code to the reader. The disadvantage lies in its considerable implementation requirement, as in both the transmitter and the receiver a transformation has to be effected. Further, the original information is not identifiable in the transmitting code vector, i.e. the code vector is not separable into information and checking symbols.

These disadvantages are avoided if the encoding and the decoding, which are only outlined here, are performed in the time domain (a more detailed description can be found, for example, in [CLARK]):

- The arithmetical operations in the Galois field, the generator polynomial of the Galois field, and the primitive field element α are the same as in the encoding and decoding in the frequency domain.
- As a first step, the information symbols are placed in the first m positions of the code vector to be transmitted.
- The encoding takes place with the aid of a code-specific generator polynomial which is constructed in such a way that its zero positions are bound to produce $2t$ consecutive zeros in the spectrum.
- If the previously generated polynomial, in which the information is stored in the first m positions, is divided modulo the generator polynomial and if the result is transferred to the last k positions of the code vector, there is a guarantee that the code vector can be divided by the generator polynomial without a remainder and that it therefore has the same roots. Thus it is a valid code vector in accordance with the definition in section 6.2.2 (its spectrum containing $2t$ consecutive zeros).
- During the decoding a so-called syndrome is initially calculated with the aid of which the key equation for $\Lambda(Z)$ is set up.
- One can solve this key equation by using Euclid's or Berlekamp's algorithms. In addition, a so-called error-evaluator polynomial $\Omega(Z)$ is generated.
- By substituting all powers of α for $\Lambda(Z)$ and $\Omega(Z)$ and by evaluating the results, one obtains the information needed about place and value of the error in the received vector.

Although the theory is somewhat more difficult to understand than that of the encoding/decoding in the frequency domain, which is mainly due to the introduction of another polynomial and the evaluation of the polynomials $\Lambda(Z)$ and $\Omega(Z)$, the implementation is much easier to realise. At the transmitting end no transformation is any longer required, and at the receiving end

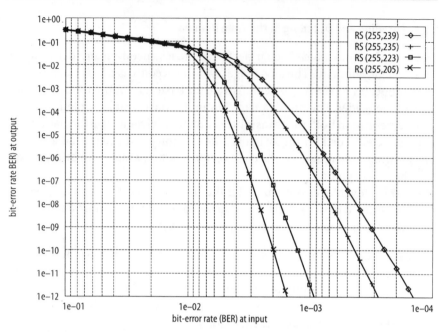

Fig. 6.9. Residual bit-error rate of various (255, 255-2t) Reed-Solomon codes

the only transformation to be done is that of the 2*t* characters of the syndrome.

6.2.6 Efficiency of the RS Code

Independently of whether the encoding or decoding takes place in the time or the frequency domain, the efficiency of both procedures with regard to their correctability is the same. Figure 6.9 shows the curve of the residual bit-error rate, i.e. the bit-error probability at the decoder output, using various RS codes as defined in $GF(2^8)$. Although the RS codes are symbol-oriented codes, the analysis of the efficiency takes bit errors into account. On the assumption that the bit errors are evenly distributed a statement can be made about the symbol-error rate on the basis of which the efficiency of an RS code can be analysed.

The efficiency of the code increases, of course, with an increase in the number of test symbols. In this way, at an input bit-error rate of $2 \cdot 10^{-3}$ the residual bit-error rate of the RS(255, 205) code is approx. $1 \cdot 10^{-10}$ – the coding gain is thus more than 10 to the power of 7 –, whereas in the case of the RS(255, 239) code at the same input bit-error rate of $2 \cdot 10^{-3}$ the output bit-error rate is $9 \cdot 10^{-4}$, the coding gain thus being only slightly greater than 0,5. The correction limit is reached when the bit-error rate at the output is greater than, or the same as, the input bit-error rate. It is in the order of $7 \cdot 10^{-3}$ at RS(255, 205) and in the order of $2 \cdot 10^{-3}$ at RS(255, 239). From this limit onwards more errors occur than the code can correct. As a consequence, additional errors occur which are caused by the corrective algorithm in the output data stream, which leads to the bit-error rate at the output of the decoder being greater than at the input.

For all DVB transmission standards a modified (shortened) RS(255, 239) code is used which makes it possible to guarantee a residual bit-error rate of approx. $1 \cdot 10^{-11}$ at an input bit-error rate of $2 \cdot 10^{-4}$ while correcting up to 8 symbol errors per block (see chapters 9 to 11).

6.3 Convolutional Codes

6.3.1 Basics of the Convolutional Codes

As already explained in section 6.1, the convolutional codes are binary codes in which the information is spread over several transmitting symbols. A convolutional code is therefore always bit-oriented. The information, consisting of individual bits, is fed into a shift register in order to be encoded. The transmitting signal is obtained by combining various taps at the shift register. Figure 6.10 shall serve as an explanation of the encoding and the technical terms used here.

Using the relatively simple convolutional encoder, as shown in figure 6.10, for each individual bit fed into the shift register, two bits are generated as output symbols. The number of input lines is $m = 1$ bit, the number of output lines is $n = 2$ bits, and the code rate $R = m/n = 1/2$. The memory, that is, the storage depth of the encoder, is defined as the number of the preceding bits which contribute to the encoding of the current bit. In this example this would be $S \cdot m = 4$, with $S = $ length of the shift register. The constraint length K

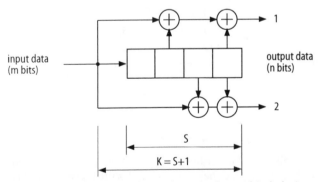

Fig. 6.10. Model construction of a convolutional code to explain the basic terms

describes the total number of bits contributing to the coding process. In the example $K = (S + 1) \cdot m = 5$. If the convolutional encoder is conceived as an automaton, as known from informatics, it can be characterised by the number of possible internal states. This number can be determined from the number of memory elements and the corresponding number of possible 1/0 combinations. The convolutional encoder shown therefore has $2^{S \cdot m} = 16$ possible states. Finally, a convolutional encoder is characterised by the number and positioning of the taps at the shift register, which are indicated by generator polynomials G the coefficients of which are 0 or 1, according to whether there was a tap at the respective position or not.[10] It is common practice to combine the coefficients as octal numbers. Polynomials or octal numbers must be indicated individually for each output branch. The characteristic parameters of this example of a convolutional encoder can be summarised as follows:

Number of input lines	m	$= 1$	
Number of output lines	n	$= 2$	
Code rate	R	$= m/n = 1/2$	
Memory	$S \cdot m$	$= 4$	
Possible states	$2^{S \cdot m}$	$= 16$	
Constraint length	K	$= (S+1) \cdot m = 5$	
Generator polynomial 1	G_1	$= 1+X^2+X^4$	(25_{OCT})
Generator polynomial 2	G_2	$= 1+X^3+X^4$	(31_{OCT})

Figure 6.11 shows a further example of a convolutional encoder which, with a number of input lines of $m = 2$, uses two shift registers with a length of $S = 3$.

The characteristic parameters of this convolutional encoder are:

Number of input lines	m	$= 2$	
Number of output lines	n	$= 3$	
Code rate	R	$= m/n = 2/3$	
Memory	$S \cdot m$	$= 6$	
Possible states	$2^{S \cdot m}$	$= 64$	
Constraint length	K	$= (S+1) \cdot m = 8$	
Generator polynomial 1	G_1	$= 1+X^2+X^3$	(15_{OCT})
Generator polynomial 2	G_2	$= 1+X^1+X^2$	(07_{OCT})
Generator polynomial 3	G_3	$= 1+X^1+X^3$	(13_{OCT})

---

[10] The LSB of the generator polynomial describes the input of the shift-register cell, whereas the MSB describes the tap at the most delayed shift-register cell. See [PROAKIS].

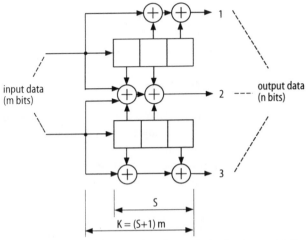

Fig. 6.11. Further model construction of a convolutional code to explain the basic terms

6.3.2 Examples of Convolutional Encoding and Decoding

6.3.2.1 Construction of a Model Encoder

In this section a very simple convolutional encoder is introduced, by means of which an example of encoding and decoding will be carried out in section 6.3.2.3. Figure 6.12 shows the construction of this encoder. It is to be noted that the input data are fed in from the right, as opposed to those in figures 6.10 and 6.11. In this way the state value of the shift register can be used, for example, to represent a state diagram (see next section).

The characteristics of this convolutional encoder are:

Number of input lines m $= 1$
Number of output lines n $= 2$
Code rate R $= m/n = 1/2$
Memory $S \cdot m = 2$
Possible states $2^{S \cdot m} = 4$

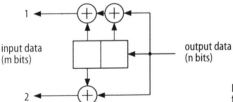

Fig. 6.12. Convolutional encoder as a basis for the following encoding and decoding example

Constraint length K $= (S+1) \cdot m = 3$
Generator polynomial 1 G_1 $= 1+X^1+X^2$ (7_{OCT})
Generator polynomial 2 G_2 $= 1+X^2$ (5_{OCT})

6.3.2.2 State Diagram and Trellis Diagram of the Model Encoder

In section 6.3.1 it was briefly mentioned that a convolutional encoder can also be considered an automaton. First of all, an automaton is characterised by the number of its internal states. As a function of the current input symbol and of its current internal state the automaton outputs one or more symbols – here bits – and changes to a new state. This can usually be represented by a state diagram as shown in figure 6.13.

The 1/0 combinations in the circles describe the state of the automata or the actual contents of the shift register. Two transition arrows diverge from each state and a further two converge to it. The first bit of the 1/0 combination at the transition arrow describes the input bit. The last two bits are the data of the outputs 1 and 2.

Commencing with the state "oo" (i.e. the shift register contains "o" in each cell), let us feed in a "1" at the input. In accordance with figure 6.12, the output signals are "11". This process is represented in the state diagram by the left-hand arrow pointing from the bottom to "o1", the designation of which is, logically, "1/11", between the states "oo" and "o1". The encoder is now in the state "o1".

Another way to document the same combinations is the trellis diagram. Here the states are plotted below one another and time-sequentially next to each other (see figure 6.14).

From every state two paths lead to a new state each, depending on whether a "1" (thin line) or a "o" (thick line) has been input. The output data is regis-tered at the transition lines. If, once again, we take state "oo" as the starting

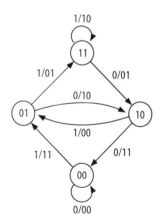

Fig. 6.13. State diagram of the model encoder

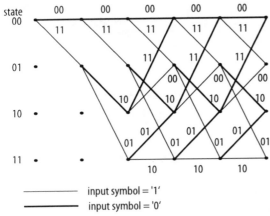

state

| 00 | 00 | 00 | 00 | 00 |

input symbol = '1'
input symbol = '0'

Fig. 6.14. Trellis diagram of the model encoder

position and read in a "1", we obtain state "01" (moving along the thin line) while reading out a "11". This corresponds to the state diagram represented in figure 6.13.

The following examples will be explained on the basis of the trellis diagram and the state diagram.

6.3.2.3 Example of Encoding with Subsequent (Viterbi) Decoding

Let the information sequence to be coded and transmitted be "1011000". In the transmission channel, let two positions be overlaid with an error, leading to the inversion of the respective bits. The transmission process can thus be summarised as follows:

Information (assumed):

1   0   1   1   0   0   0

Coded bit sequence (to be transmitted):

11   10   00   01   01   11   00

Error vector (assumed):

01   00   10   00   00   00   00

Received sequence:

10   10   10   01   01   11   00

The total Viterbi decoding is explained on the basis of the trellis diagrams in figure 6.15 together with the state diagram in figure 6.13. As opposed to the encoding, in the decoding process the input and the output of the state diagram now have to be exchanged.

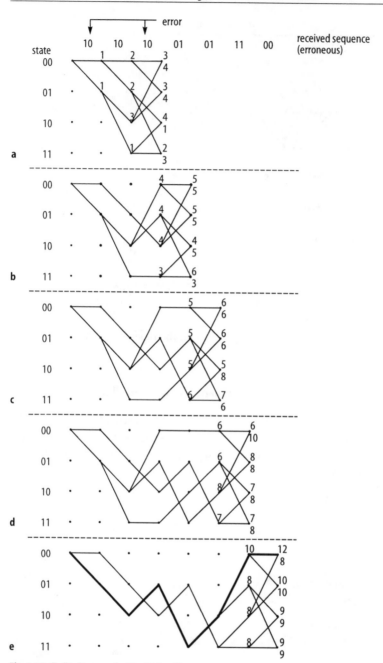

Fig. 6.15. Trellis diagrams for Viterbi decoding

The decoding takes place as follows:

(1) The decoder is in the state "oo" and receives the bit sequence "10". As shown in the state diagram, the encoder could not have generated this bit sequence, because commencing with the state "oo" there are only two alternative possibilities:
  – Sending a "oo", and keeping the state "oo" (in the trellis diagram in 6.15a the top left-hand horizontal transition line). The decoder 'knows' that in this case only one correct bit was received. As the sum of the correct bits a "1" is recorded in the transition path.
  – Sending a "11", and transition to state "o1" (in the trellis diagram in 6.15a the diagonal line from the top left-hand side). The decoder 'knows' also in this case that only one correct bit was received. A "1" is again recorded as the sum of the correct bits.

(2) The decoder again receives the bit sequence "10".
  – Starting from state "oo" when decoding the second input bit sequence, the reception of a "oo" and the retention of state "oo" is again expected or, alternatively, the reception of a "11" and a transition to state "o1". This step, too, would produce one incorrect bit. In each of the two cases 2 correct bits have now been recognised (out of 4 bits received in the meantime).
  – Starting from state "o1", the reception of a "o1" with a transition to state "11" and one correct bit still in the sum, or the reception of a "10" with a transition to state "10" and hence a total of 3 correct bits is expected.

At this point we can introduce the term 'metric'. In the present case the metric Δ denotes the sum of the correctly received bits, while following a chosen path through the trellis diagram. The larger the metric, the higher the probability that the path through the trellis diagram will correspond to the path that has been followed through the encoder. For a mathematical description of the metric see section 6.3.3, (6.24) to (6.26).

(3) After the third bit sequence "10" has been received (see figure 6.15a), all possible transitions between the states have been analysed, and the input bit sequence has been compared with the expected reception values, the case occurs that in each state position two transitions converge. The principle of the Viterbi decoder (see [VITERBI]) is now to choose precisely that transition out of two which has the larger metric or, in other words, to delete the transition with the lower metric and, as the case may require, the preceding transitions as well, since this is the less probable path through the trellis diagram. The result can be seen in the trellis diagram shown in figure 6.15b. Should two transitions with the same metric converge, then a transition can be chosen at random, as from this position backwards there

is no unambiguous decision possible. For a proof of the above the reader is referred to [VITERBI].

(4)–(6) By processing the following received bit sequences "01", "01" and "11" and successively deleting the transitions with lower metrics a path through the trellis diagram emerges whose metric is larger than that of the other paths (figures 6.15b to 6.15d).

(7) The last bit sequence received is "00". The transitions to be chosen can be seen in figure 6.15e.

The path of the metric $\Delta = 12$ through the trellis diagram is now the most probable (drawn thickly in figure 6.15e). By retracing, one finally obtains the most probable state sequence, and with the aid of the state diagram one obtains the original data sequence:

Most probable state sequence:

| 00 | 01 | 10 | 01 | 11 | 10 | 00 | 00 |

Decoded and corrected received sequence:

| 11 | 10 | 00 | 01 | 01 | 11 | 00 |

Information sequence:

| 1 | 0 | 1 | 1 | 0 | 0 | 0 |

The errors in the received sequence have thus been corrected. In addition, by comparing the highest metric with the number of bits received, a statement can be made about the number of errors which occurred (in this case: $14 - 12 = 2$ errors) and therefore also about the actual state of the transmission channel.

6.3.3 Hard Decision and Soft Decision

Figure 6.16 shows the generalised probability density of an originally binary received signal which has been transmitted over a disturbed channel. Either the symbol $x_i = 0$ or the symbol $x_i = 1$ has been transmitted. Due to noise in the transmission channel no signal with discrete states has been received but,

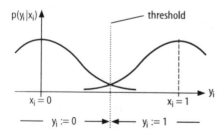

Fig. 6.16. Probability density of the received signal and hard-decision threshold

instead, a signal y_i with a wide range of values and with the conditional probability density function $p(y_i|x_i)$.

In hard-decision decoding, the range of y_i is divided by a threshold. For all reception values of y_i which are above the threshold, $y_i := 1$ is assumed. For all others $y_i := 0$ is valid. The metric δ_i for the comparison of two bits in hard-decision decoding is therefore

$$\delta_i = \begin{cases} 1 & \text{for } x_i = y_i \\ 0 & \text{for } x_i \neq y_i \end{cases}. \tag{6.24}$$

In soft-decision decoding, on the other hand, several interim states for y_i are evaluated. The range of values for y_i is divided by several thresholds and quantised as represented, for example, in figure 6.17 which depicts a 3-bit soft decision.

The range of the metric for the comparison of two bits x_i and y_i in soft decision extends from 0 to 1 in 8 steps and is identified by the parameter d_i (cf. figure 6.17):

$$\delta_i = \begin{cases} d_i & \text{for } x_i = 1 \\ 1 - d_i & \text{for } x_i = 0 \end{cases}. \tag{6.25}$$

The metric Δ of a path through the trellis diagram results in both cases (hard decision and soft decision) from the sum of the metrics δ_i of the bit comparisons, as conveyed in the description of the decoding.

$$\Delta = \sum_i \delta_i \tag{6.26}$$

We can see that in soft-decision decoding non-integral metrics are also possible. This leads to a far more accurate estimation of the probability that a chosen path through the trellis diagram is correct. The typical coding advantage, which is to be gained by the use of soft decision in the decoder, is in the range of 2 dB (cf. figure 9.11, chapter 9).

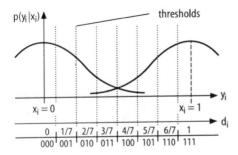

Fig. 6.17. Probability density of the received signal and soft-decision thresholds

6.3.4 Puncturing of Convolutional Codes

One of the disadvantages of convolutional codes is the low code rate, which is $R = 1/2$ in the encoder used here. This means that there are twice as many bits transferred than the actual information would require, or to state it differently: the data stream contains 50% redundancy. By puncturing, the code rate can be increased, which of course increases the correction requirement. As an example, we will again revert to the information sequence from section 6.3.2.3. If the model encoder is used, this information sequence results in the coded data of rate $R = 1/2$. In that now every third bit of the coded data sequence is not transmitted, or is punctured, the code rate increases to $R = 3/4$. For the purpose of explaining the puncturing, an error-free transmission is here assumed. The decoder tries to reconstruct the original coded data of rate $R = 1/2$ by assuming an X (= "don't care") after each second symbol received. When calculating the metric for the Viterbi decoding each symbol received is used in accordance with the rules of soft decision in section 6.3.3, whilst for "don't care" a metric of $\delta_x = 0{,}5$ is assumed. The principle of encoding, including puncturing and decoding, is recapitulated in the following table.

Information sequence (assumption):

1	0	1	1	0	0	0

Coded data, rate 1/2:

11	10	00	01	01	11	00

Puncturing to rate 3/4:

11	$_1 0$	0_0	01	$0 1$	1_1	00

Data transferred (here error-free):

11	00		01	11		00

Reconstruction for decoding:

11	Xo	oX	01	X1	1X	00

Metric δ_i with soft decision:

$\delta_1\,\delta_2$	$0{,}5\,\delta_4$	$\delta_5\,0{,}5$	$\delta_7\,\delta_8$	$0{,}5\,\delta_{10}$	$\delta_{11}\,0{,}5$	$\delta_{13}\,\delta_{14}$

6.3.5 Performance of Convolutional Codes

In figure 6.18 the residual bit-error rate of convolutional codes of rate $R = 1/2$ is plotted as a function of E_b/N_0. E_b/N_0 is defined as the energy E_b, which is transmitted per bit, divided by the noise-power density N_0 of the white Gaussian noise with which the signal on the transmission channel is overlaid.[11]

---

[11] For further explanations see chapter 7.

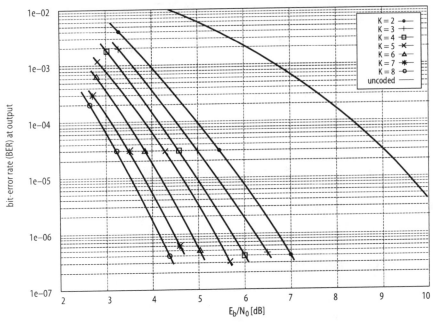

Fig. 6.18. Residual bit-error rate of convolutional codes of rate R = 1/2 in QSPK modulation

Furthermore, we assume a transmission with QPSK modulation (see chapter 7), the parameter K describing the constraint length of the code used.

The performance of the error correction increases – as expected – with increased constraint length of the code used. The more preceding information bits are used for the encoding of an output symbol, i.e. the wider the input information is "smeared", the more reliable the data are, as a result of the decoding. So, for example, with an E_b/N_0 of 5 dB at the input of a decoder, the residual bit-error rate at a constraint length of $K = 2$ is less than two powers of ten better than in the uncoded case, while for $K = 6$ a residual bit-error rate can be achieved which is more than four powers of ten better than in the uncoded case.

For the DVB standard a convolutional code of rate $R = 1/2$ with a constraint length of $K = 7$ is used. In this way it is possible, while having an E_b/N_0 of only 3.2 dB[12], to achieve a bit-error rate of less than $2 \cdot 10^{-4}$ at the output of the decoder, this ratio corresponding to the maximum permissible bit-error rate at the input of the RS decoder, so that finally a bit-error rate at the output of the RS decoder of less than $1 \cdot 10^{-11}$ is obtained (cf. section 6.2.6).

---

[12] This value corresponds to the one that was theoretically determined. No allowance for an implementation margin has been made here (cf. section 9.5.2).

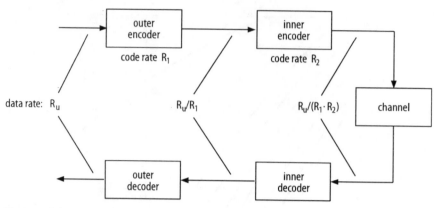

Fig. 6.19. Code concatenation

6.4 Code Concatenation

By the concatenation of various codes it is possible to increase the performance of the error correction; however, this also increases the transmission data rate. Figure 6.19 shows the concatenation of two codes.

Let the information data rate to be transmitted be R_u and the first-used code have the code rate R_1. This code is called outer code, as it is either right at the beginning of the transmission path or right at the end. The second code has the code rate R_2 and is called the inner code. Therefore the data rate to be transmitted is $R_u/(R_1 \cdot R_2)$. The advantage of code concatenation is explained in the following section by means of the concatenation of two block codes.

6.4.1 Block-Code Concatenation

The simplest example of a concatenation of different codes is the concatenation of block codes (see figure 6.20).

The outer code here is an RS code, which adds k redundant symbols to the m information symbols. The frame length of the outer code is therefore $n = m + k$. Each individual symbol is comprised of m' bits. The inner code, in this case a Hamming code, now adds to the m' bits of each symbol of the outer code k' redundant bits for error protection. An individual bit error will now be corrected by the inner code in the decoder so that the corresponding symbol is correct for the outer code, i.e. the outer decoder is not burdened by bit errors. On the assumption that the inner code can only correct one bit error, up to n bit errors can occur per frame of the outer code without the outer decoder having to correct any symbol error. A prerequisite for this is that each of the n bit errors fits perfectly within a frame of the inner code vector. Without

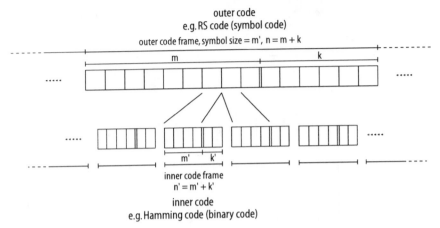

Fig. 6.20. Block-code concatenation

the inner code there would have been a decoding breakdown in the outer decoder. Should there be more than one bit error per inner code frame, then
these errors can no longer be corrected by the inner code. In the outer code
there is, however, only one erroneous symbol in each code vector which has to
be corrected. A code concatenation, as shown in figure 6.20, thus combines
the advantages of the individual codes, viz, reliable correctability of individual bit errors in the inner code and reliable correctability of short burst errors
in the outer code.

6.4.2 Interleaving

In order to correct long burst errors in addition to bit errors and short burst
errors, an interleaver is inserted between the outer and the inner code (see
figure 6.21).

The interleaver supplies no additional error correction code, it merely ini-
tiates a rearrangement of the symbols generated by the outer code. The principle is shown in figure 6.22.

The symbols generated by the outer encoder (here 1 symbol = 1 byte) are
read into the storage matrix of the block interleaver line by line. Thereafter
the matrix is read out column by column and the symbols are fed individually
to the inner encoder. After interleaving, two adjacent symbols generated by
the outer encoder are separated from each other by exactly the number of

Fig. 6.21. Interleaving during channel coding

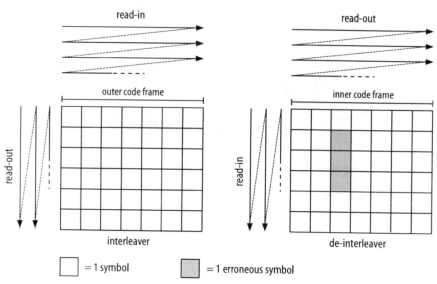

Fig. 6.22. Operating principle of a block interleaver

symbols that can be found in a column of the matrix. This number is the main parameter of an interleaver and is called interleaving depth I. In the example of figure 6.22 I equals 6.

At the receiver the output symbols of the inner decoder – which may be overlaid by errors – are deposited column by column in the matrix of the de-interleaver. A lengthy burst error, which could not be corrected by the inner decoder, will show in one of the columns (shaded in figure 6.22). Due to the line-by-line read-out of the matrix, there will only be one symbol error per frame of the outer code. Without an interleaver/de-interleaver all symbol errors of a burst would be within one frame of the outer code. A decoding breakdown would be the result.

The disadvantages of the block interleaver are that periodic disturbances which invalidate the same symbol in each column would result in a decoding breakdown of the outer decoder. This would then place the erroneous symbols in one line of the de-interleaver. Apart from this, the block interleaver has a relatively high requirement for storage capacity, which is also a disadvantage; another disadvantage being the two-dimensional synchronisation, i.e. not only the beginnings of the outer code vectors have to be found, but also the first outer code frame which stands in the first line of the de-interleaver. These disadvantages are avoided by the utilisation of the convolutional interleaver, according to [FORNEY], as shown in figure 6.23.

The convolutional interleaver consists of $(I - 1)$ shift registers with the length $M, 2M, \ldots, (I - 1)M$ and corresponding multiplexers and demultiplex-

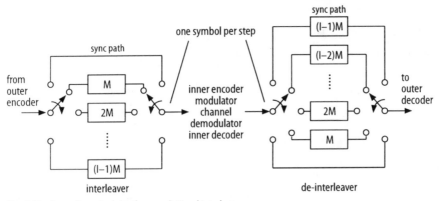

Fig. 6.23. Operating principle of a convolutional interleaver

ers (represented here as switches), each of which connects a shift register with an input or an output. As before, I represents the interleaving depth and M the so-called base delay. $M = n/I$ applies if n, as before, represents the frame length of the outer code. With each step the multiplexer and the demultiplexer switch on to the next respective input or output. The next symbol is read into the shift register currently connected to the input, and a another symbol is picked up from the output of that shift register. When the top path in the interleaver, the sync path, happens to be active, the input is connected directly to the output. This ensures that, between adjacent symbols at the input, $M \cdot I$ further symbols are transmitted.

The de-interleaver is constructed in such a way as to ensure that the non-delayed symbols of the interleaver are delayed to a maximum. Hence, the total delay is $M \cdot (I - 1) \cdot I$ for all symbols. At the beginning of each outer frame all multiplexers and demultiplexers (in this case switches) must be in the initial position, i.e. only a synchronisation of one level is required. The interleaving depth I must therefore be a whole-number divisor of the frame length n of the outer code (see above).

6.4.3 Error Correction in DVB

For the transmission of digital TV over satellite in accordance with [ETS 421] (see chapter 9) and via terrestrial transmission networks in accordance with [ETS 744] (see chapter 11) an RS code is concatenated with a convolutional code by means of an interleaver (see figure 6.24).

The RS code is based on the Galois field $GF(2^8)$ and therefore has a symbol size of 8 bits. An RS code (255,239) was chosen which processes a data block of 239 symbols and can correct up to 8 symbol errors by calculating 16 redundant correction symbols. As an MPEG-2 transport packet is 188 bytes long

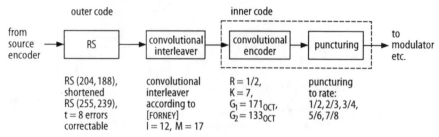

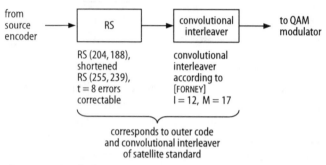

Fig. 6.24. Coding for forward error correction in the transmission of digital TV via satellite and terrestrial network

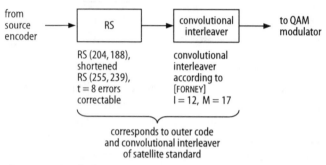

Fig. 6.25. Coding for forward error correction in the transmission of digital TV via cable

(see chapter 5), the code was shortened, i.e. the first 51 information bytes were set to zero and not transmitted at all. In this way an RS code (204, 188) is generated.

After the outer code a convolutional interleaver according to [FORNEY] is used. The interleaving depth is $I = 12$. From the frame length of the outer code with $n = 204$ the base delay results as $M = n/I = 17$.

Finally, a convolutional code is applied to the interleaved symbols. Its rate is $R = 1/2$, the constraint length is $K = 7$. The taps at the shift register are described by the two generators $G_1 = 171_{OCT}$ and $G_2 = 133_{OCT}$. Optionally, a puncturing to the rates $R = 2/3, 3/4, 5/6$, and $7/8$ is possible. Coding for error correction by transmission over cable in accordance with [ETS 429] is similar to the aforementioned coding (see figure 6.25), only the convolutional code is not required as the signal-to-noise performance in the cable channel is very much better than in the satellite channel.

A detailed description of the systems in their entirety as well as their performance can be found in chapters 9, 10 and 11.

6.5 Further Reading

To those who wish to read more extensively in the theory of codes for error correction and error recognition and who intend to consolidate their understanding of the concepts introduced here, the following works are recommended for further reading. A very basic and comprehensible overview of block and convolutional codes is given in [SWEENEY], whilst the block codes, from the basics to the deeper theory, are treated in [FURRER]. There is a standard work by G.C. Clark and J.B. Cain [CLARK], which, among other things, discusses aspects of implementation, compares various algorithms and contains performance graphs (residual bit-error rate versus signal-to-noise power). The original publication by S. Reed and G. Solomon on the code of the same name is to be found in [REED], that by A.J. Viterbi in [VITERBI]. [FORNEY] discusses the convolutional interleaver for the channel with overlaid burst errors, while [HAGENAUER] concentrates on the theory of the puncturing of convolutional codes.

Symbols in Chapter 6

$A(Z)$	polynomial of the Galois field in the frequency domain
A_i, A_l	i^{th}, l^{th} coefficient of $A(Z)$
$a(X)$	polynomial of the Galois field in the time domain
a_i, a_l	i^{th}, l^{th} coefficient of $a(X)$
$C(Z)$	code vector (polynomial) transformed from $c(X)$
$C_i,$	i^{th} coefficient of $C(Z)$
$c(X)$	code vector (polynomial) to be transmitted
c_i	i^{th} coefficient of $c(X)$
d_i	quantised input signal for soft decision of ith bit comparison
$E(Z)$	error vector (polynomial) transformed from $e(X)$
E_b	energy/bit
E_i	i^{th} coefficient of $E(Z)$
$e(X)$	error-vector polynomial
e_i	i^{th} coefficient of $e(X)$
$GF(q)$	Galois field of size q
G_i	generator polynomial at output i of a convolutional encoder
$g(x)$	generator polynomial of Galois field
I	interleaving depth
i	running variable, integer
K	constraint length of a convolutional code
k	number of correction symbols of a block code/outer block code
k'	number of correction symbols of an inner block code
l	running variable, integer
M	base delay of the convolutional interleaver
m	number of information symbols in a block code/outer block code or input bits in the convolutional code
m'	number of information symbols in the inner block code
N	$2^w - 1$
N_0	noise-power density

n number of code symbols in a block code/outer block code or input bits in the convolutional code

n' number of code symbols in an inner block code

$p(y_i \mid x_i)$ probability that the symbol x_i is sent and the symbol y_i is received

q size (number of elements) of a Galois field

R code rate

$R(Z)$ received vector (polynomial) transformed from $r(X)$

R_1, R_2 rate of the outer code, rate of the inner code

R_i i^{th} coefficient of $R(Z)$

R_u useful data rate

$r(X)$ received vector (polynomial)

S length of a shift register

t number of correctable errors in RS coding

w exponent of 2 in the definition of q

X argument of polynomials defined in time domain of GF

x polynomial argument in GF

x_i i^{th} symbol sent

y_i i^{th} reference symbol, to be compared to the symbol received

Z argument of polynomials defined in frequency domain of GF

α primitive element of Galois field

α, β_i elements of the Galois field

Δ metric of a trellis path

δ_i metric of the ith bit comparison

$\Lambda(Z)$ transformed polynomial from $\lambda(X)$

Λ_i i^{th} coefficient of $\Lambda(Z)$

$\lambda(X)$ error-locator polynomial

λ_i i^{th} coefficient of $\lambda(X)$

$\Omega(Z)$ error-evaluator polynomial

7 Digital Modulation Techniques

MPEG source coding, which achieves a data reduction in audio and video signals, has been discussed in chapters 3 and 4. As explained in chapter 5, the various elementary streams are combined in the MPEG transport multiplexer to form a single data stream. This is followed by a coding of the data stream in which redundant signal portions are inserted (see chapter 6). The entire processing of the baseband signals can take place in a computer (or a digital circuit) in which the data are available as a sequence of numerical values. For these values to be transmitted on a channel they have to be converted into genuine data signals. The signals are output sequentially by an interface, synchronisation being provided by an internal processing clock. Each information bit possesses finite energy, which will be referred to as bit energy E_b. The methods by which data signals can be adapted to the respective transmission channel are the subject of the following section.

7.1 NRZ Baseband Signal

The physical shape of the signal which is mainly used for the processing of digital signals in the baseband is called non-return-to-zero (NRZ). An ideal signal is usually described as a sequence of weighted Dirac pulses [LÜKE 1]. The data are read out at the output interface of the computer, where they adhere to a rigid time-slot pattern nT_B. Their shaping into an NRZ pulse is achieved by a hold unit, which holds the value of the information for the period of the timing pulse T_B so that each ideal Dirac pulse is overlaid with a rectangular pulse. Figure 7.1 depicts an NRZ signal with the period T_B. The amplitude assumes the value of A when state 1 is to be transmitted, and the value of 0 when state 0 is to be transmitted.

If the signals, having assumed this form, are to be used for data transmission, it is important to know their spectral properties. The power spectral density (PSD or spectrum, for short) of the NRZ signal can be computed on the assumption that the signal states consist of a statistically independent sequence of values occurring with equal frequency. The PSD consists of one discrete portion PSD_{dis} and one continuous portion PSD_{con} [MORGENST]:

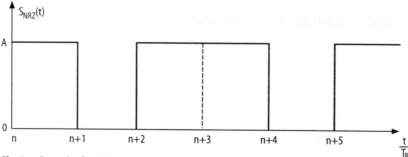

Fig. 7.1. Example of an NRZ signal with a symbol period of T_B

$$PSD_{NRZ}(\omega) = PSD_{dis} + PSD_{con} = \frac{A^2}{4}\delta(\omega) + \frac{A^2}{4}T_B \cdot si^2\left(\frac{\omega T_B}{2}\right) \tag{7.1}$$

The discrete component in the first part of (7.1) is limited to a single Dirac pulse at a radian frequency $\omega = 0$. It represents the continuous portion of the signal, which can also be seen in fig 7.1. The rectangular pulse shape of the time-domain signal generates the continuous portion in the PSD. This rapidly diminishes with an increase in frequency. Ideally, in the case of infinitely steep signal edges there would result an infinitely wide PSD. This is plotted in fig. 7.2.

The post-filter effect of the rectangular pulse shape is conspicuous [SCHÖNFD 1]. However, as a rule this does not suffice for a real transmission. In most cases the signal spectrum has to be limited to a finite value by an additional filter to account for the finite bandwidth of the channel. The suppression of the high-frequency portions of the signal results in a widening of the impulses in the time domain and causes temporal crosstalk between the individual impulses. This behaviour is called "intersymbol interference (ISI)". In accordance with the first Nyquist criterion it is sufficient for the signal to have no ISI at the sampling points nT_B. The course of a single impulse traversing the sampling points nT_B, should have the following sampling values:

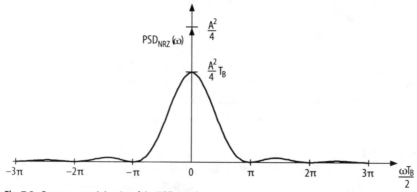

Fig. 7.2. Power spectral density of the NRZ signal

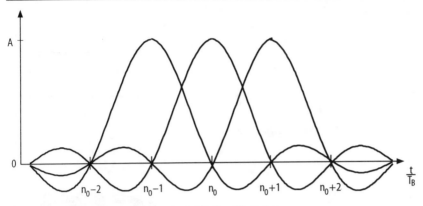

Fig. 7.3. Interference-free transmission due to ideal band limitation

$$s_{NRZ}(t-nT_B) = A \quad \text{for} \quad n = n_0$$
$$s_{NRZ}(t-nT_B) = 0 \quad \text{for} \quad n \neq n_0 \tag{7.2}$$

This first Nyquist criterion is satisfied, among others, by the si-shaped course of an impulse, which can be generated by an ideal band limitation at half the clock frequency. The cut-off frequency is referred to as the Nyquist frequency f_N. Figure 7.3 shows the curve represented by three consecutive si-impulses. The resulting signal is interference-free at the sampling points because the si-functions of all consecutive impulses are zero at these points.

In real systems, however, the ideal sampling accuracy required cannot be maintained and, moreover, the filters used always have finitely steep filter slopes. Nyquist proved in his second theorem that, with a cosine-wave frequency response $H(f)$, the ISI on both consecutive impulses is limited. For each of the individual impulses, an additional requirement – apart from the first Nyquist criterion – is that the amplitude values which are exactly midway between the two consecutive samples have half the basic impulse amplitude (in this case 0.5 A) [NYQUIST]:

$$s_{NRZ}(t-nT_B) = \frac{A}{2} \quad n = n_0 \pm \frac{1}{2}, \tag{7.3}$$

A cosine-wave frequency response is shown in figure 7.4 a. Figure 7.4 b shows the corresponding eye diagram. The amplitude values can be clearly seen from (7.2) and (7.3). The disadvantage resulting from adherence to the second Nyquist criterion is due to the doubling of the channel bandwidth required for the transmission. For this reason the cosine-wave drop in frequency response is generally reduced to a narrower transition band which is symmetrical to the Nyquist frequency. The resulting frequency responses correspond to (7.4), with the roll-off factor indicating the steepness of the filter slope and the ensuing extension of the frequency range. The roll-off factor can assume a

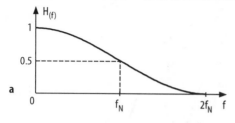
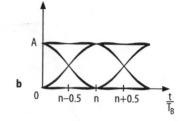

Fig. 7.4. a Frequency response, **b** Eye diagrams for $\alpha = 1$

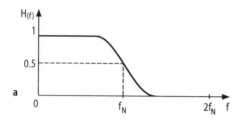
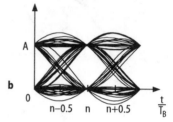

Fig. 7.5. a Frequency response, **b** Eye diagrams for $\alpha = 0{,}35$ (e.g. DVB satellite)

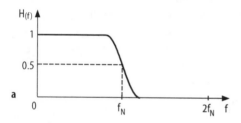
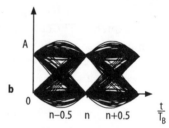

Fig. 7.6. a Frequency response, **b** Eye diagrams for $\alpha = 0{,}15$ (e.g. DVB cable)

value between 0 and 1. The frequency responses have a so-called "raised co-sine" characteristic. Two examples with differing roll-off factors are shown in figures 7.5 a and 7.6 a, which are used for the transmission of DVB signals over satellite (see chapter 9) and cable (see chapter 10). It can be clearly seen by the eye diagrams in figures 7.5 b and 7.6 b that the first Nyquist criterion is satisfied. The second Nyquist criterion, however, is infringed, as not all individual impulses have the required amplitude values at $(n \pm 0.5)T_B$ according to (7.3).

$$H(f) = 1 \qquad\qquad \text{for } |f| < f_N (1-\alpha)$$

$$H(f) = \tfrac{1}{2} + \tfrac{1}{2} \sin\left[\frac{\pi}{2f_N} \left(\frac{f_N - |f|}{\alpha} \right) \right] \quad \text{for } f_N(1-\alpha) \le |f| \le f_N(1+\alpha) \qquad (7.4)$$

$$H(f) = 0 \qquad\qquad \text{for } |f| > f_N(1+\alpha)$$

In practice the required Nyquist signal frequency response is generally obtained by two subfilters, one connected after the other. The first subfilter is at the transmitter output. It limits the signal spectrum to $f_N (1 + \alpha)$ and shapes the impulse. In the following it will be assumed that the data signal is disturbed by additive white Gaussian noise (AWGN). The subfilter at the receiver input has the task of minimising the noise power which was added during the transmission and of maximising the signal-to-noise ratio S/N at the sampling point. An arrangement in which the impulse response of the filter is identical to the impulse response of the signal transmitted is called a matched filter [NORTH]. The effect of the noise signal no longer depends on the wave-form of the useful signal, but only on the average energy E_b which is required per bit. If the useful signal has no redundancy (e.g. no error-protection portion), this results in a signal-to-noise ratio of

NOTE

$$\frac{S}{N} = \frac{E_b}{N_0} \tag{7.5}$$

[LÜKE 1, KAMMEYER] at the optimal sampling point at the filter output.

The change in the digital signal brought about by the AWGN signal causes the decoder to make a false decision if the noise amplitude exceeds a given decision threshold. In accordance with the above assumption the noise signal has a Gaussian amplitude-density distribution [GAUSS], also called "normal distribution"

$$f(a) = \frac{1}{\sigma_N \sqrt{2\pi}} \cdot e^{-\frac{(a-\mu)^2}{2\sigma_N^2}} \tag{7.6}$$

with

amplitude a,

mean value $\mu = 0$ for average-free noise, $\qquad$ (7.6a)

variance $\sigma_N^2 = N$.

The curve of this function has been tabulated in various publications (e.g. [BRONST]). Figure 7.7 shows the superposition of the two possible values (A and 0) in a symmetrical channel, each having the same Gaussian amplitude-density function. The optimal decision threshold is marked by a broken line and is to be found at the amplitude $\alpha_{thresh} = 0.5\ A$.

The probability of a detection error can be calculated for each state from the area below that part of the curve which lies beyond the decision threshold. The primitive of this function necessary for this purpose is not available in closed form. Hence, the so-called error function was introduced for the path integral:

$$erf(x) = \frac{2}{\sqrt{\pi}} \int_0^x e^{-z^2} dz. \tag{7.7}$$

The probability of a false decision is given by the so-called complementary error function (erfc):

$$erfc(x) = 1 - erf(x). \tag{7.8}$$

A connection between the maximum amplitude of data signal A and noise power N results from the substitution of the variable z from (7.7) and by making use of (7.6.a):

$$z = \frac{a - a_{thresh}}{\sqrt{2\sigma_N^2}} \Rightarrow x = \frac{A - \frac{A}{2}}{\sqrt{2N}} = \frac{A}{\sqrt{8N}}. \tag{7.9}$$

In various publications (e.g. [LÜKE 1]) an instantaneous power $S_a = A^2$ is defined which is used as a reference quantity for computing the bit-error rate. In this section an average signal power S shall serve as reference quantity (as, for instance, in [KAMMEYER]). Assuming that the transmitted states and the corresponding amplitude values A and o occur, on average, with equal frequency, the average signal power S can be computed as follows:

$$S = \tfrac{1}{2}(A^2 + 0) = \frac{A^2}{2} \Rightarrow A = \sqrt{2S}. \tag{7.10}$$

Substituting (7.10) into (7.9) and making use of (7.5) we obtain:

$$x = \sqrt{\frac{E_b}{4N_0}}. \tag{7.11}$$

As the signal in accordance with figure 7.7 can only be invalidated in one direction, the complementary error function must be multiplied by the value 0.5. Thus the bit-error rate (BER) for the baseband transmission of a unipolar NRZ signal results in the required dependency of E_b/N_0:

$$BER_{unipol} = \tfrac{1}{2} erfc\left(\sqrt{\frac{E_b}{4N_0}}\right). \tag{7.12}$$

In a bipolar baseband transmission, state o is mapped onto the value $-A$ by a negative excursion of the amplitude, which causes the mean signal power to increase to

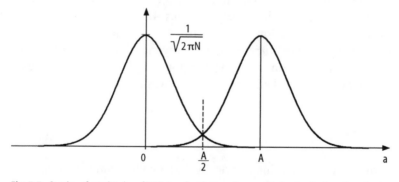

Fig. 7.7. Overlay of amplitudes of NRZ signal with Gaussian amplitude-density function of noise

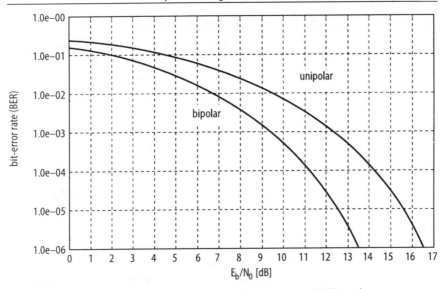

Fig. 7.8. Bit-error rate of a unipolar and a bipolar baseband transmission with NRZ signals

$$S = \tfrac{1}{2}(A^2 + A^2) = A^2 \Rightarrow A = \sqrt{S} \,. \tag{7.13}$$

The decision threshold α_{thresh} from (7.9) is shifted to the DC potential. If these changes are taken into account, the result is the following bit-error rate for a bipolar baseband transmission:

$$BER_{bipol} = \tfrac{1}{2}erfc\left(\sqrt{\frac{E_b}{2N_o}}\right). \tag{7.14}$$

Both bit-error rates are shown in figure 7.8 as functions of E_b/N_o.

Although in bipolar transmission, the distance of the amplitude range between the decision threshold and the two possible amplitude values has doubled as opposed to the unipolar transmission, this only results in a noise-ratio improvement by 3 dB owing to the duplication of the average signal power S.

In those publications which define the above-mentioned instantaneous power S_a as reference quantity, this gain increases to 6 dB.

7.2 Principles of the Digital Modulation of a Sinusoidal Carrier Signal

There are various transmission systems for the distribution of broadcast signals. These include satellite systems, cable networks and terrestrial channels, which are generally organised as frequency multiplex systems. For the transmission of user data the available frequency ranges are divided into system-specific channel spacings, a fact which requires an adaptation of the baseband signals to the particular channel conditions. This takes place by the modulation of a carrier signal. Heed has to be taken that the power of the data signal be concentrated as much as possible at the required position in the frequency spectrum. For this reason sinusoidal carrier signals are chosen:

$$s_{\cos}(t) = \text{amplitude} \cdot \cos(2\pi \cdot \text{frequency} \cdot t + \text{phase}). \qquad (7.15)$$

Before the signals are converted, band limitation and impulse pre-shaping are usually performed at the transmitter (see section 7.1). After transmission in the high-frequency range, the signal must be down-converted to the baseband and filtered by the matched filter at the receiver. Usually each signal which passes through one of the above-mentioned transmission media would be processed in accordance with section 7.1 prior to modulation and subsequent to demodulation. The previous considerations concerning the baseband transmission with regard to the filters used in the transmitter and the receiver therefore retain their validity for the transmission in a real band-pass channel. In this case, too, the useful signal is overlaid with an AWGN signal with noise-power density N_0 which is constant over the whole frequency range. An optimal receiver synchronously demodulates the signal and then calculates the real part of the baseband signal [KAMMEYER]. After the signal has passed through the matched filter, sampling takes place with the double Nyquist frequency $2f_N$, which provides discrete amplitude values A' as a result.

The above findings can be translated into the high-frequency range before demodulation. On account of the band-pass character of the modulated signals we obtain, for each transmitted state, a defined amplitude and a reference state (or rather, a reference frequency for frequency shift keying) of the carrier signal used. If these states are plotted in a complex plane, the Euclidean distance d is a measure of the resistance of this method against disturbances. This complex amplitude plane is described in the literature as a constellation diagram. Moreover, to facilitate a comparison between the various modulation techniques it is common practice to scale to the value of 1, resp. $\sqrt{2}$, the smallest signal amplitude which differs from zero.

By analogy with what has been said about baseband transmission, the probability of an erroneous band-pass transmission can be defined by substituting into (7.9) the Euclidean distance d for the transmitted amplitude A. On this assumption the noise power N is equal to the real effective noise power. Various publications, such as [LÜKE 1], describe the modulation techniques on the basis of this approach. However, in our subsequent discussion the total power which is added to the noise signal during transmission over a real transmission channel will be referred to as N in accordance with [KAMMEYER]. This means that on account of the above-mentioned calculation of the real part of the signal in the receiver only half the noise power N is effective, so that, in addition, 0.5 N needs to be substituted for σ_N^2 in (7.9). Therefore the result for the argument of the complementary error function is:

$$x = \frac{d}{\sqrt{4N}}.\qquad(7.16)$$

According to (7.15) it is possible to modulate the information to be transmitted onto the amplitude (amplitude shift keying – ASK), the frequency (frequency shift keying – FSK), the phase (phase shift keying – PSK), or onto a combination (e.g. quadrature amplitude modulation – QAM). In this way symbols are generated whose feature-carrying information (e.g. the amplitude in ASK) can assume different states. As an introduction, the principles of the digital modulation techniques will be explained by means of the three basic methods. This will be followed by a discussion of the techniques which are intended for the transmission of DVB signals.

7.2.1 Amplitude Shift Keying (2-ASK)

In 2-ASK the unipolar NRZ signal, discussed in section 7.1, will be fed to a modulator. After its connection with the carrier, the NRZ signal switches the modulator on when its state is 1, and switches it off when a 0 is to be transmitted. Because of this keying mode the 2-ASK is also referred to as "on-off keying". It is very easy to implement by means of a switch or, more generally, by a multiplier. The block diagram of the modulator and the wave-forms of the time-domain signals are shown in figure 7.9.

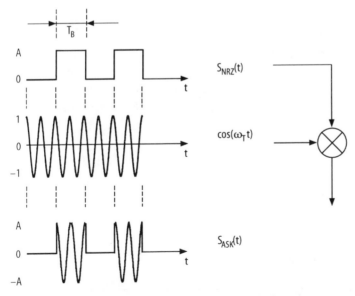

Fig. 7.9. Block diagram of the 2-ASK modulator and time graph of signals

It can be clearly seen that the information transmitted is contained in the envelope of the signal. This phenomenon is typical of amplitude modulation and also known from analogue technology.

Under the same conditions as in section 7.1 (the same occurrence probability and statistical independence of the two states) the temporal multiplication of the unipolar NRZ signal by the sinusoidal carrier results in a PSD which arises from a convolution of the PSDs of the two individual signals and consists of one discrete portion and one continuous portion [JOHANN].

$$PSD_{2-ASK}(\omega) = PSD_{dis}(\omega) + PSD_{con}(\omega)$$

$$= \frac{A^2}{16}\left[\delta(\omega_T + \omega) + \delta(\omega_T - \omega)\right] \tag{7.17}$$

$$+ \frac{A^2}{16}T_B\left[si^2\left((\omega_T + \omega)\frac{T_B}{2}\right) + si^2\left((\omega_T - \omega)\frac{T_B}{2}\right)\right]$$

Because of the discrete signal portion of the carrier frequency ω_T a synchronous demodulation in the receiver is possible without additional measures being taken.

Figure 7.10 shows the one-dimensional constellation of 2-ASK. Also plotted is the square root of the mean signal power S so that the relationship between S and the Euclidean distance d can be read off. This results in:

$$d = \sqrt{2S} . \tag{7.18}$$

From (7.18), (7.16) and (7.5), we obtain the bit-error rate

$$BER_{2-ASK} = \tfrac{1}{2}erfc\left(\sqrt{\frac{E_b}{2N_0}}\right). \tag{7.19}$$

With 2-ASK it is not imperative to perform a synchronous demodulation. Figure 7.10 depicts the decision threshold by means of a broken circular line with a radius of half the amplitude. As can be seen, it is also possible to perform a demodulation of the envelope, independently of the phase angle transmitted.

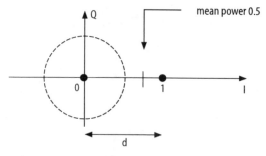

Fig. 7.10. Constellation diagram of 2-ASK and of the amplitude value which corresponds to the mean ASK power

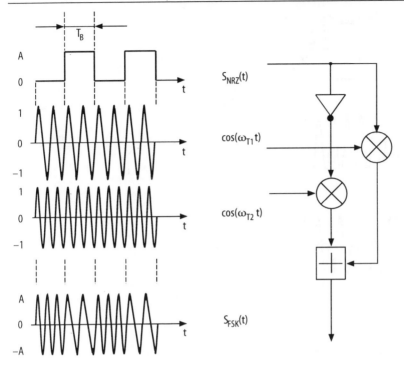

Fig. 7.11. Block diagram of the 2-FSK modulator and time graph of signals

7.2.2 Frequency Shift Keying (2-FSK)

The findings from section 7.2.1 can be used to explain the 2-FSK technique if the 2-FSK modulator is conceived as two 2-ASK modulators connected in parallel (cf. figure 7.9). The two multiplier outputs are interconnected by means of an adder. The unipolar NRZ signal discussed in section 7.1 is read into one of the multiplier inputs. The inverted unipolar NRZ signal is fed into the other multiplier as an input signal. The carrier signals used for the modulation have two different frequencies. They are sampled by the input signals whose wave-forms are in inverse relation to each other, so that the symbol occurring at the output of the adder always differs from zero. This process results in the wave-form of a time-domain signal with a constant envelope. The block diagram of the 2-FSK modulator and the wave-forms are shown in figure 7.11.

The PSD of the 2-FSK technique results from the addition of two $PSD_{2\text{-}ASK}$, which, owing to the different carrier frequencies, are spectrally offset against each other.

A more thorough treatment of the 2-FSK technique is given, for example, in [JOHANN].

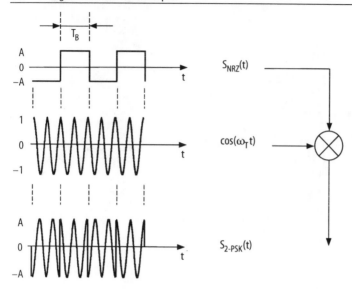

Fig. 7.12. Block diagram of the 2-PSK modulator and time graph of signals

7.2.3 Phase Shift Keying (2-PSK)

A further possibility of transmitting information is to change the phase angle of the carrier signal by 180°. By means of the relation

$$\cos(\omega t + \pi) = -\cos(\omega t) \tag{7.20}$$

the 2-PSK can also be interpreted as an ASK, although the amplitude is not switched on and off, as is the case with amplitude keying, but is inverted. When using a 2-ASK modulator, as shown in figure 7.9, this inversion of the carrier can be achieved very simply by feeding a bipolar NRZ signal to the input instead of a unipolar NRZ signal. The block diagram of the modulator and the corresponding wave-forms of the time-domain signals can be seen in figure 7.12.

The constant envelope of the 2-PSK can be clearly seen in the lower part of figure 7.12. Furthermore, a comparison with the wave-form of 2-ASK (see bottom part of figure 7.9) shows that intervals during which the signal is zero, such as occur in 2-ASK, can be filled by inversion or by a 180° phase shift of the carrier. This causes the power of the 2-PSK signal to double compared with the mean power of a 2-ASK signal. On the assumption that the two symbol states are statistically independently distributed and occur, on average, with equal frequency, the carrier, temporally averaged out over numerous symbol states, must be absent from the spectrum. Hence this is a modulation technique with carrier suppression. The PSD of 2-PSK can be obtained from the

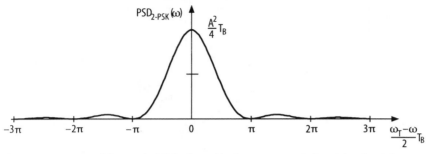

Fig. 7.13. Power spectral density of 2-PSK for the positive frequency range in the neigbourhood of $\omega = \omega_T$

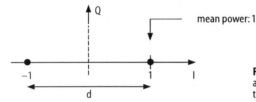

Fig. 7.14. Constellation diagram of 2-PSK and of the amplitude value which corresponds to the mean PSK power

PSD of 2-ASK (see (7.17)) by suppressing the discrete component and, on the basis of the above-mentioned power balance, quadrupling the continuous component. According to [JOHANN]:

$$PSD_{2-PSK}(\omega) = \frac{A^2}{4} T_B \left[si^2 \left((\omega_T + \omega) \frac{T_B}{2} \right) + si^2 \left((\omega_T - \omega) \frac{T_B}{2} \right) \right]. \qquad (7.21)$$

The PSD in figure 7.13 represents the positive frequency range with $\omega = \omega_T$ if $T \gg 2\pi/T_B$.

The one-dimensional constellation of 2-PSK is shown in figure 7.14. The decision threshold, whose course is orthogonal to the transmitted in-phase component, passes exactly through the origin of the co-ordinates. The relation between the Euclidean distance d and the mean signal power S is calculated as

$$d = 2\sqrt{S}. \qquad (7.22)$$

As the signal has the same power in both states a symbol decision can only be carried out after its synchronous demodulation. As opposed to 2-ASK, the required carrier has to be recovered by non-linear signal processing. The Costas loop is a method which is often used for the synchronous demodulation of the PSK signals [COSTAS].

The bit-error rate for 2-PSK can be calculated from (7.22), (7.16) and (7.5):

$$BER_{2-PSK} = \frac{1}{2} erfc \left(\sqrt{\frac{E_b}{N_o}} \right). \qquad (7.23)$$

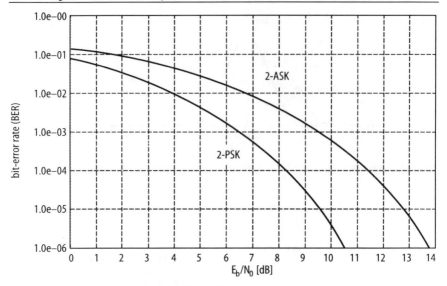

Fig. 7.15. Bit-error rates for 2-PSK and 2-ASK transmission

The bit-error rate BER for 2-ASK and 2-PSK is given in figure 7.15 as a function of E_b/N_0. A doubling of the mean signal power results in an improvement in the signal-to-noise ratio of 3 dB for 2-PSK as compared to 2-ASK. Several publications do not relate the bit-error rate of 2-ASK to the total signal power, but take the AC component of the signal power as a reference value [MÄUSL]. In this way the bit-error rate curve of 2-ASK is shifted by 6 dB to higher signal-to-noise ratios.

7.3 Quadrature Phase Shift Keying (QPSK)

The QPSK technique offers the possibility of simultaneously transmitting two bits per symbol. This doubles the spectral efficiency (controlling the number of bits that can be transmitted per second per required bandwidth) as opposed to the techniques dealt with up to now. The serial data stream of the NRZ signal is first split up into two parallel paths by means of a demultiplexer. This process is shown in figure 7.16 for eight consecutive bits. After the data stream has been transformed into two parallel ones, each two bits can be processed simultaneously. These dibits are assigned the function of a complex symbol with a real and an imaginary part. The wave-forms of the signals are referred to as $Re\{s_{Dibit}(t)\}$ or $Im\{s_{Dibit}(t)\}$. The duration which is available for the processing of a complex symbol is referred to as symbol duration T_S. It is double the bit duration T_B in the case of QPSK.

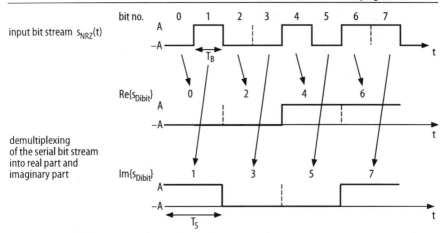

Fig. 7.16. Generation of dibits from the serial data stream

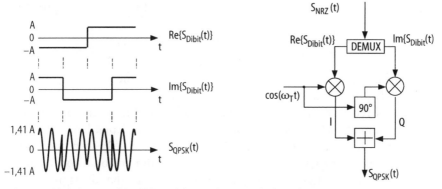

Fig. 7.17. Block diagram of the QPSK modulator and time graph of signals

Figure 7.17 shows the block diagram of the QPSK modulator. The demultiplexer at the input divides the serial data stream into two parallel paths. The scalar components are fed to one multiplier each. Two sinusoidal signals, having the same frequency ω_T and being phase-shifted against each other by 90°, serve as carriers. The two signal parts are then added together. This type of signal processing, as shown in (7.24), can be expressed by a complex multiplication of both quantities with the subsequent calculation of the real part of the signal. Hence the QPSK signal at the output of the modulator can be computed as:

$$s_{QPSK}(t) = Re\left\{ s_{Dibit}(t) \cdot e^{-j\omega_T t} \right\}$$

$$= Re\{s_{Dibit}(t)\} \cdot \cos(\omega_T t) - Im\{s_{Dibit}(t)\} \cdot \sin(\omega_T t). \tag{7.24}$$

From figure 7.17 it is apparent that the QPSK signal results from the addition of two 2-PSK signals. The spectrum of the QPSK signal is the outcome of the

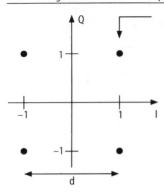

Fig. 7.18. Constellation diagram of QPSK and of the amplitude value which corresponds to the mean QPSK power

addition of the two 2-PSK spectra. The symbol duration T_S must be used in (7.21) instead of the bit duration T_B. Moreover, the signal power is doubled by the addition of the two 2-PSK signals, which leads to the following expression:

$$PSD_{QPSK}(\omega) = \frac{A^2}{2} T_S \left[si^2 \left((\omega_T + \omega) \frac{T_S}{2} \right) + si^2 \left((\omega_T - \omega) \frac{T_S}{2} \right) \right]. \qquad (7.25)$$

By substituting the relation $T_S = 2T_B$ one realises that the required transmission bandwidth is halved as against that in 2-PSK.

A QPSK receiver separates the input signal into its in-phase and quadrature components, where each can be regarded as an independent 2-PSK signal, as mentioned in the last paragraph. This can be clearly seen once again when we look at the QPSK signal in the now two-dimensional constellation diagram depicted in figure 7.18. Moreover, it can be seen that the signal power S is equally distributed between the two components. Half the signal power $S/2$ is effective for both bit decisions, in the direction of the in-phase component I as well as in the direction of the quadrature component Q. The resulting amplitude value equals half the Euclidean distance, which is illustrated in 7.18.

$$d = 2 \cdot \sqrt{\frac{S}{2}} = \sqrt{2S} \qquad (7.26)$$

The bit-error rate results from the substitution of the Euclidean distance into (7.16). Using (7.5), we have to substitute the bit energy E_b by the symbol energy E_s since the principle adopted in section 7.1 for the optimisation of the signal-to-noise ratio by the matched filter in the receiver no longer refers to individual bits but to the transmitted symbols.

$$\frac{S}{N} = \frac{E_s}{N_o} \qquad (7.27)$$

E_s corresponds to the sum of the energy of the two signal components E_b which are required per bit:

$$E_s = 2E_b .$$
(7.28)

For the bit-error rate we get:

$$BER_{QPSK} = \tfrac{1}{2} erfc\left(\sqrt{\frac{E_b}{N_0}}\right).$$
(7.29)

As both the signal power S and the noise power N are distributed between the two components I and Q the same conditions apply to each signal component as in 2-PSK. Therefore it is not surprising that the bit-error rate of a QPSK signal is the same as the mean value of the bit-error rates of its two signal parts and that it is thus identical with the bit-error rate of a 2-PSK signal. This statement is valid as long as the bit-error rate is given as a function of E_b/N_0.

Apart from the interpretation of the QPSK signal as a combination of two 2-PSK signals, QPSK can be conceived as a quadrature amplitude modulation (QAM) with 4 states. That both interpretations are valid is shown by a comparison between the bit-error rate curves in figure 7.15 (for 2-PSK) and figure 7.23 (for 4-QAM).

Apart from the basic QPSK variant discussed above, other special forms are discussed in the literature [e.g. KAMMEYER, JOHANN, MÄUSL, PROAKIS] in detail (i.e. Offset-QPSK, DPSK).

QPSK has practical applications in the transmission of digital signals over satellite channels (see chapter 9) and in connection with the OFDM technique (see section 7.6) in terrestrial transmissions (see chapter 11).

7.4 Higher-level Amplitude Shift Keying (ASK) and Vestigial-Sideband Modulation (VSB)

In television engineering, analogue vestigial-sideband modulation is to be found in PAL colour coding as well as in television transmissions. Today's terrestrial transmission channels which are used for radio broadcasting are designed to accommodate VSB, just as are the cable networks, given that the digital variant of the vestigial-sideband modulation only differs from the analogue one in that a finite number of discrete amplitude states have to be transmitted. Therefore there are economic advantages in using the existing systems for digital VSB without any changes having to be made.

The serial data stream at the input of the modulator is first combined to form data words of width m and then allocated to the M possible discrete symbol states by means of a table, where

$$M = 2^m .$$
(7.30)

The required VSB signal shape can be generated using ASK. As the modulated signal contains no phase information it will suffice for only one of the side-

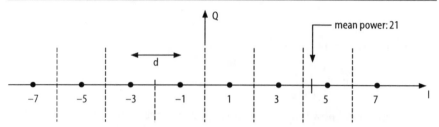

Fig. 7.19. Constellation diagram of 8-level ASK and of the amplitude value which corresponds to the mean ASK power

bands to be transmitted. After the modulation the second sideband is therefore almost completely suppressed by post-filtering. This, however, produces negative side effects. For example, the coherent carrier signal in the receiver which is required for the synchronous demodulation cannot be recovered from the transmitted VSB signal. Hence the modulator must not completely suppress the carrier. The residual carrier becomes noticeable in the one-dimensional constellation of the VSB by an additive DC component and thus increases the mean signal power. The one-dimensional constellation of an eight-level ASK, as shown in figure 7.19, differs from the eight-level VSB only by the missing DC component. As a rule, the individual standardised amplitude levels A_l' are indicated in the literature in accordance with (7.31):

$$A_l' = 2l - 1 - M \quad \text{for} \quad l = 1, 2, \dots, M.\tag{7.31}$$

At this point it should be mentioned that the ASK technique for the value $M = 2$, as described in this section, does not correspond with the 2-ASK technique introduced in section 7.2.1, because in higher-level ASK the amplitude values of (7.31) are modulated symmetrically to 0. The ASK with the value $M = 2$ corresponds to the 2-PSK introduced in section 7.2.3.

The mean power S of the ASK signal is computed from the expected value of the squares of all possible signal amplitudes A_l', assuming that all amplitudes occur with the same probability [JOHANN, PROAKIS]. For 8-ASK, the amplitude value corresponding to the mean power S is plotted in figure 7.19.

$$S = E\left\{A_l'^2\right\} = \frac{2}{M} \sum_{l=1}^{M/2} (2l-1)^2 = \frac{M^2 - 1}{3}\tag{7.32}$$

The basic relationship between S and the Euclidean distance d follows from figure 7.19, using (7.32):

$$\frac{d}{\sqrt{S}} = \frac{2}{\sqrt{\frac{M^2-1}{3}}} \Rightarrow d = \sqrt{4 \cdot \frac{3}{M^2-1} \cdot S}.\tag{7.33}$$

For higher-level modulation the bit-error rate (BER) can be estimated via the symbol-error rate (SER). The symbol-error rate indicates the probability of an error occurring in the detection of a transmitted symbol. For the computation of the SER an additional correction factor needs to be taken into account, which can be explained as follows: higher-level ASK techniques, as opposed to the techniques discussed before, have more than two amplitude levels. As a consequence, each level has two direct neighbours and therefore two decision thresholds beyond which it may be disturbed and found to be erroneous. Hence, the error probability increases by the factor of 2. The two outer points in the constellation which indicate the two peaks are the exceptions. Since the SER constitutes the mean value of all error probabilities of the M individual states, the correction factor is computed as follows:

$$c_{cor} = 2 \cdot \frac{M-1}{M}. \tag{7.34}$$

Hence, the M-level ASK has an SER of

$$SER_{ASK} = \frac{M-1}{M} \, erfc \left(\sqrt{\frac{3}{M^2-1} \cdot \frac{E_s}{N_o}} \right). \tag{7.35}$$

The transition from one double-sideband ASK signal to a signal modulated in accordance with the VSB technique is explained by the following simplified model:

(1) As already discussed, for the transmission almost the whole of one sideband is suppressed. The receiver filter, as a result, has only half the noise bandwidth. This results in a correction of the signal-to-noise ratio of 3 dB to the advantage of the VSB signal.

(2) In the case of synchronous demodulation, the correlated amplitudes of the upper and the lower sideband in the sampling point of an ASK signal are vectorially superimposed. As the vector representing the lower sideband is missing in a VSB transmission, the Euclidean distance in the constellation is reduced by half. This means that the signal-to-noise ratio is reduced by 6 dB as compared with the signal-to-noise ratio of an ASK signal.

(3) Furthermore, the disadvantage of additionally required power for the transmission of the coherent carrier signal becomes noticeable. In the U.S., where the Grand Alliance favours VSB as the transmission technique for cable networks and terrestrial channels (see chapter 1), a power increase of 0.3 dB has been recommended [GRALLI].

The total balance corresponds to a deterioration in the signal-to-noise ratio by a factor of approx. 2.14, which corresponds to a logarithmic value of 3.3 dB. For the transition from ASK to VSB a reduction in the required transmission bandwidth is exchanged for a loss in immunity to interference. By the suppression of one sideband the spectral efficiency is increased from the ideal $m/2$ bit/s per Hz for ASK to the ideal m bit/s per Hz for VSB [SCHÖPS].

The symbol-error rate for VSB can be estimated from (7.35) by using the above factor of 2.14:

$$SER_{VSB} = \frac{M-1}{M} erfc\left(\sqrt{\frac{3}{2.14(M^2-1)} \cdot \frac{E_s}{N_0}}\right). \tag{7.36}$$

The dependence on the bit energy E_b is computed using the following relation (see also (7.28) for QPSK):

$$E_s = m \cdot E_b . \tag{7.37}$$

The average bit-error rate can be obtained by taking into account the fact that when a symbol is transmitted correctly all m bits are transmitted error-free as well. On the assumption that, firstly, all m bits are statistically independent of each other and, secondly, that they are erroneously detected with the same probability, the following approximation applies:

$$1-SER \cong (1-BER)^m \Rightarrow BER \cong 1-(1-SER)^{\frac{1}{m}} \tag{7.38}$$

Therefore the bit-error rate of an ASK signal and of a VSB signal can be expressed as follows:

$$BER_{ASK} = 1 - \left[1 - \frac{M-1}{M} erfc\left(\sqrt{\frac{3}{(M^2-1)} \cdot \frac{m \cdot E_b}{N_0}}\right)\right]^{\frac{1}{m}} \tag{7.39}$$

$$BER_{VSB} = 1 - \left[1 - \frac{M-1}{M} erfc\left(\sqrt{\frac{3}{2.14(M^2-1)} \cdot \frac{m \cdot E_b}{N_0}}\right)\right]^{\frac{1}{m}}. \tag{7.40}$$

The bit-error rate for various VSB techniques is shown in figure 7.20. The curves for ASK can be easily obtained by shifting the VSB curves by a value of 3.3 dB to lesser E_b/N_0 values.

Different characteristics of VSB are to be found in table 7.1. It can be seen that the mean signal power increases by approx. 6 dB per level towards higher constellations, which results in a reduction of the Euclidean distance d and in

Table 7.1. Characteristics of the VSB techniques

Number of symbol states and bits per symbol	Mean signal power	Ratio between peak and mean signal power	Dynamic range
$M, m = ld\ (M)$	$10 \log\left(\frac{M^2-1}{3}\right) + 0.3$ dB	$10 \log\left(3\frac{(M-1)^2}{M^2-1}\right)$	$20 \log (M-1)$
2, 1	0.3 dB	0.0 dB	0.0 dB
4, 2	7.3 dB	2.6 dB	9.5 dB
8, 3	13.5 dB	3.7 dB	16.9 dB
16, 4	19.6 dB	4.2 dB	23.5 dB

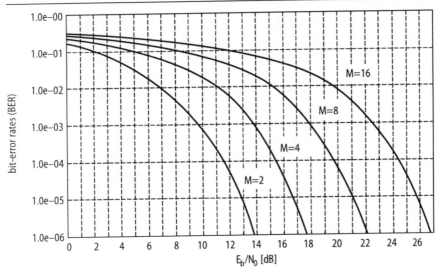

Fig. 7.20. Bit-error rates of the VSB techniques for various M

the correlated reduction of the interference immunity. The ratio of the maximum signal power, which is represented by the outer constellation points, to the mean signal power S is important, if, for example, possible overdrive effects of amplifiers need to be avoided. The minimum signal power is always equal to 1 according to (7.31). A signal dynamics of 0 dB signifies a constant envelope.

7.5 Digital Quadrature Amplitude Modulation (QAM)

In section 7.3 it was shown how by demultiplexing the serial data stream into two parallel branches a complex symbol can be generated. In section 7.4 m bits were combined to a higher-level symbol. By both techniques the spectral efficiency can be increased. Hence it is reasonable to combine the two techniques in order to obtain a further improvement.

In digital QAM m bits are combined and mapped onto a complex symbol word consisting of a real part S_{real} and an imaginary part S_{imag}. In figure 7.21 the mapper is shown at the input of the modulator. In contrast to figure 7.17, the mapper does not necessarily generate binary signals any more but, depending on the desired QAM, also multilevel ones. In QPSK the symbol word represents the dibit. The combination of the m bits results in the symbol duration T_S increasing proportionally to the bit duration. The actual modulation process that follows is identical to the technique described in section 7.3. The real part of the QAM symbol word modulates a cosine-wave carrier signal in the in-phase branch, while the imaginary part of the QAM symbol modulates a sine-wave carrier of the same frequency in the quadrature branch of the modulator. By this process two carrier-frequency oscillations are generated which can be conceived as independent ASK signals. Once more the similarity to the QPSK signal, which is composed of two 2-PSK signals, is easily seen. The required QAM signal results after adding the two components. The wave-forms of the time-domain signals are shown in figure 7.21 by taking 16-QAM as an example. It can be seen that the envelope of the QAM signal is not constant. Therefore, the possibility of carrying out an envelope demodulation in the demodulator cannot be precluded at this stage. Against this, the fact has to be borne in mind that there are various symbol states which have the same amplitude but different phase angles. For this reason only a synchronous demodulation is required. This can be demonstrated by using the two-dimensional constellation shown in figure 7.22 which takes a 64-QAM as an example. The first of the four quadrants contains 16 of the 64 possible amplitude and phase values. The scaling follows the general practice in the literature:

$$A'_{I,k} = 2k - 1 - \sqrt{M} \qquad \text{for} \quad k = 1, 2, \ldots, \sqrt{M}$$
$$A'_{Q,l} = 2l - 1 - \sqrt{M} \qquad \text{for} \quad l = 1, 2, \ldots, \sqrt{M}. \tag{7.41}$$

Using (7.41), the amplitude values marked along the negative axes correspond to the amplitude of the in-phase component $A'_{I,k}$ and the amplitude of the quadrature component $A'_{Q,l}$ respectively. The signal powers in the fourth quadrant result from the points of the vectors. If all constellation points occur with the same frequency, the mean power of the transmitted signal is obtained from the expected value of the squares of all amplitude levels, as expressed by the following equation:

$$S = E\left\{A'^2_{k,l}\right\} = \frac{1}{M} \sum_{k=1}^{\sqrt{M}} \sum_{l=1}^{\sqrt{M}} \left(A'^2_{I,k} + Q'^2_{Q,l}\right) = 2\frac{M-1}{3}. \tag{7.42}$$

Applying the above example to 64-QAM and using $M = 64$, the scaled mean power results in 42. This value is plotted in the fourth quadrant of figure 7.22.

The second quadrant in figure 7.22 shows the phase values for the various states. For convenience, these values have been so arranged as to mirror symmetrically those in the first quadrant, so that the zero phase angle results in the absolute angle of 180°. In a mathematically negative sense of rotation the phase angle increases. The actual values for the second quadrant can be computed by subtracting the given angles from 180°.

The third quadrant contains the optimal decision thresholds.

The four points which are nearest to the origin of the system of coordinates in figure 7.22 correspond to the QPSK values. Therefore QPSK can also be referred to as 4-QAM. The transmission of 16 amplitude and phase

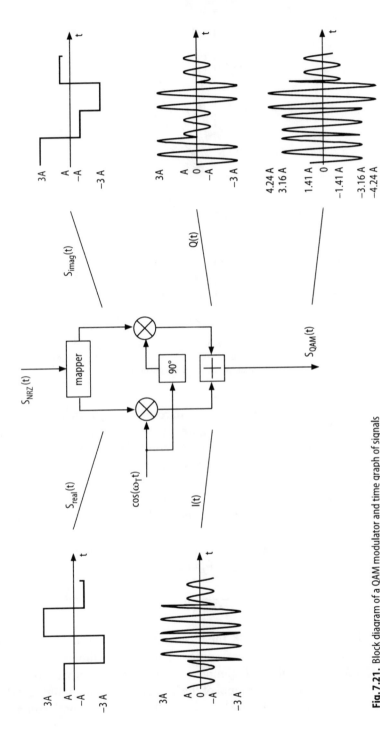

Fig. 7.21. Block diagram of a QAM modulator and time graph of signals

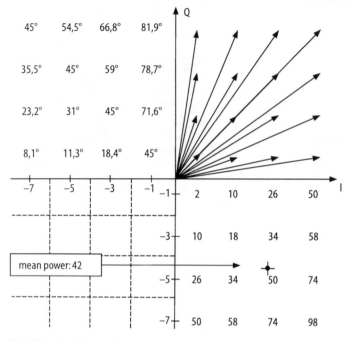

Fig. 7.22. Constellation of the 64-QAM with varying parameters:
1st quadrant: vector diagram, 2nd quadrant: relative phase angles, 3rd quadrant: decision thresholds,
4th quadrant: power of the individual symbol states and mean power

values, in accordance with a 16-QAM, results in a constellation which can be described by the 16 values in figure 7.22 which are nearest to the origin.

The special cases of those QAM constellations which result from the transmission of an odd number of m are not discussed in this section. There are various approaches to arranging the amplitude and phase values in the respective constellation. Some of the more frequently discussed solutions are called cross constellations and are explained, among others, in [LEE].

On the basis of (7.42), the general relation between the Euclidean distance d and S can be deduced from figure 7.22.

$$\frac{d}{\sqrt{S}} = \frac{2}{\sqrt{2\frac{M-1}{3}}} \Rightarrow d = \sqrt{4\frac{3}{2(M-1)}S} \tag{7.43}$$

Similar to higher-level ASK the bit-error rate is obtained by computing the SER. According to the above findings an M-QAM symbol consists of two separate $\sqrt{M}$-ASK symbols. A complex QAM symbol is only correctly decoded when not only the ASK symbol of the in-phase component but also that of the quadrature component have been correctly detected. As both components are statistically independent of each other, this results in the following SER for the M-QAM technique:

$$SER_{QAM} = 1 - \left(1 - SER_{\sqrt{M}-ASK}\right)^2 = 2SER_{\sqrt{M}-ASK} - SER^2_{\sqrt{M}-ASK} . \qquad (7.44)$$

From (7.35), by substituting the number of states M with $\sqrt{M}$, one obtains the error rate for one of two QAM components (in-phase or quadrature). Further, it is to be noted that each component has only half the QAM signal power, so that in the denominator of the argument the factor 2 has to be added:

$$SER_{I,Q} = \frac{\sqrt{M}-1}{\sqrt{M}} \, erfc\left(\sqrt{\frac{3}{2(M-1)} \cdot \frac{E_s}{N_o}}\right). \qquad (7.45)$$

The argument of the complementary error function in (7.45) can be computed by substituting the Euclidean distance d from (7.43) and the matched-filter condition from (7.27) into (7.16).

The expression of the symbol-error rate of an M-QAM symbol follows from (7.44) and (7.45):

$$SER_{QAM} = 2\frac{\sqrt{M}-1}{\sqrt{M}} \, erfc\left(\sqrt{\frac{3}{2(M-1)} \cdot \frac{E_s}{N_o}}\right)$$
$$- \left[\frac{\sqrt{M}-1}{\sqrt{M}} \, erfc\left(\sqrt{\frac{3}{2(M-1)} \cdot \frac{E_s}{N_o}}\right)\right]^2 . \qquad (7.46)$$

After conversion of the symbol energy E_s into the bit energy E_b (see (7.37)) the bit-error rate results in

$$BER_{QAM} = 1 - \left[1 - \left[2\frac{\sqrt{M}-1}{\sqrt{M}} \, erfc\left(\sqrt{\frac{3}{2(M-1)} \cdot \frac{m \cdot E_b}{N_o}}\right)\right.\right.$$
$$\left.\left. - \left[\frac{\sqrt{M}-1}{\sqrt{M}} \, erfc\left(\sqrt{\frac{3}{2(M-1)} \cdot \frac{m \cdot E_b}{N_o}}\right)\right]^2\right]\right]^{\frac{1}{m}} \qquad (7.47)$$

in accordance with (7.38).

The bit-error rates for various M are plotted in figure 7.23 as a function of E_b/N_o.

In the representation of the two-dimensional constellations of a 64-QAM signal, which are shown as "snapshots" in figure 7.24, it can be seen how the ideal positions of the transmitted values are invalidated by superimposed noise. Each position represents one sampled complex amplitude value. With $E_b/N_o = 12$ dB, the points, seemingly uncoordinated, appear to be distributed over the whole constellation. With $E_b/N_o = 17$ dB the constellation pattern

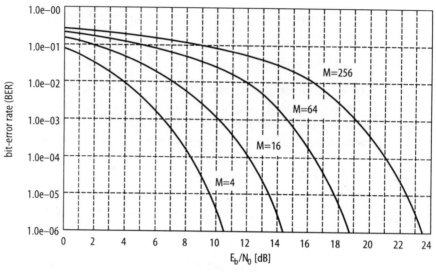

Fig. 7.23. Bit-error rates of the QAM techniques for various M

$E_b/N_0 = 12$ dB $E_b/N_0 = 17$ dB $E_b/N_0 = 22$ dB

Fig. 7.24. Constellations of a 64-QAM with $E_b/N_0 = 12$ dB, $E_b/N_0 = 17$ dB and $E_b/N_0 = 22$ dB

transmitted can already be clearly distinguished. From figure 7.23 it can be deduced that the signal now has a bit-error rate lower than 10^{-4}. When $E_b/N_0 = 22$ dB the bit-error rate drops below 10^{-6}.

Several characteristics of the QAM technique have been compiled in table 7.2. The mean power increases by about 3 dB per level. A comparison with table 7.1 shows that in the VSB technique, the one-dimensional expansion of the constellation results in an average power increase of approx. 6 dB per level. The ratio between maximum power – represented by the outer points in the constellation – and mean power is identical with the values of the VSB technique, as is the signal dynamics. However, here the values increase, with each level, by the same amount as in the VSB technique, therefore increasing the requirements for the level control of the amplifier used [FRIEDR].

Table 7.2. Characteristics of the QAM techniques

Number of symbol states and bits per symbol $M, m = ld\,(M)$	Mean signal power $10 \log\left(2\frac{M-1}{3}\right)$	Ratio between peak and mean signal power $10 \log\left(3\frac{(\sqrt{M}-1)^2}{M-1}\right)$	Dynamic range $20 \log(\sqrt{M}-1)$
4, 2	3.0 dB	0.0 dB	0.0 dB
8, 3	6.7 dB see [PROAKIS]		
16, 4	10.0 dB	2.6 dB	9.5 dB
32, 5	13.2 dB see [PROAKIS]		
64, 6	16.2 dB	3.7 dB	16.9 dB
128, 7	19.2 dB see [PROAKIS]		
256, 8	22.3 dB	4.2 dB	23.5 dB

The QAM techniques will be used for the transmission of DVB signals in cable networks (see chapter 10) and also for terrestrial transmission (see chapter 11) in connection with the OFDM technology (see section 7.6).

7.6 Orthogonal Frequency Division Multiplex (OFDM)

A terrestrial broadcasting channel differs from a satellite transmission link or a cable transmission channel in that it is prone to multipath propagation. Reflections of the transmitted signal from obstacles such as buildings or mountains are superimposed asynchronously on the directly received signal. These reflected signals are, of course, time-delayed and can cause harmful interference. If the delay time of the echo signals approaches the symbol duration of the transmitted signal, this circumstance results in a selective behaviour of the frequency response [KAMMEYER]. Such a scenario is shown in figure 7.25, where it is aggravated by additional interference from a co-channel.

The individual echoes which successively arrive at the receiver, vary in amplitude and delay time. By superimposing themselves on the main signal they cause fluctuations in the complex channel transfer function. A characteristic value of such fading channels is given by the ratio of the directly received signal power to the total of the power of all echo signals.

By using suitable equalisers it is possible to compensate for the distortions in the frequency domain. The time delays of the various echoes, however, often by far exceed the symbol duration. This means that a corresponding number of adjacent symbols affect each other. A filter for ISI reduction must therefore be of a high order, which makes its implementation very expensive.

It is possible to minimise the number of symbols affecting each other by lengthening the duration of the symbol transmitted. This can be done by the

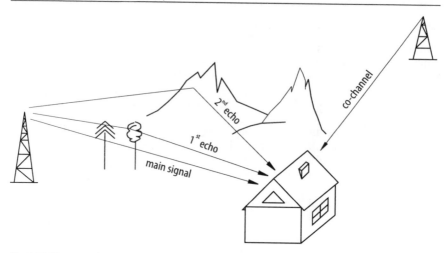

Fig. 7.25. Transmission scenario of a possible multipath reception in one channel with co-channel transmitter

parallel transmission of several symbols. If, for instance, the information to be transmitted is simultaneously modulated onto 1000 symbols of different carrier frequencies, then for each individual symbol (in the following referred to as subsymbol) there is a time slot available, which, before changing to parallel transmission, was allotted to all the 1000 sequentially transmitted symbols together. One of the basic criteria in communication engineering implies that the values of the bandwidth and the transmission time of an information can vary as a function of each other. In this way the frequency range required for the transmission of an individual subsymbol is reduced by the corresponding value. The total bandwidth of all subsymbols remains almost constant as compared with the bandwidth when the single-carrier technique is used. Figure 7.26 shows the block diagram of a multicarrier system at the encoder. As with QAM or QPSK, first of all m bits from the serial data

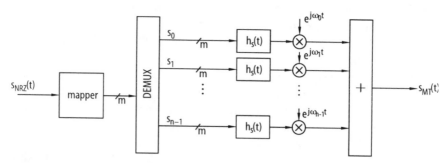

Fig. 7.26. Block diagram of a multicarrier system

stream are combined and mapped into complex subsymbols. Second, after a combination of the subsymbols to obtain the desired number to be transmitted in parallel, a serial-parallel conversion takes place. At the transmitting end, in each of the parallel branches there is a pre-filter (with the impulse response $h_t(t)$) as well as a modulator which modulates the respective subsymbol to the desired frequency position. Thereafter, these signals, which are often referred to as subcarriers, are added to a multicarrier symbol. The signal path which exists in the case of a multicarrier symbol combining n subsymbols s_k (as illustrated in figure 7.26) can be expressed as:

$$s_{MT}(t) = \sum_{k=0}^{n-1} s_k \cdot h_t(t) \cdot e^{j\omega_k t} . \tag{7.48}$$

As each individual subsymbol s_k can be modulated in amplitude as well as in phase, the multicarrier technique now also uses the third parameter – the frequency – for transmitting the information.

The complexity of this procedure increases to the extent that the number of parallel branches augments as a consequence of an increase in the number of necessary filters and modulators, so that this kind of implementation can very quickly lead to high costs.

A special case of the multicarrier technique is the OFDM system. For this a prerequisite is that all subcarrier frequencies ω_k be orthogonal to each other:

$$\omega_k \equiv 2\pi k f_o \quad \text{where} \quad k = 0, 1, 2, \ldots, n-1$$
$$\text{and} \quad f_o \text{ as base frequency} . \tag{7.49}$$

In this case the parallel connection of the modulators in figure 7.27 exactly follows the rule for computing the inverse discrete Fourier transform (IDFT), with a subsequent frequency conversion of the entire signal.

The IDFT is a block-oriented processing algorithm [OPPENHM]. It is necessary for a predetermined number of subsymbols to be available simultaneously at the inputs of the IDFT unit. For this reason the sequentially received data are temporarily stored, until the required number of subsymbols for parallel transmission have accumulated, and are then read out in parallel. Figure 7.28 demonstrates by a simple example the principle of the signal processing within the subsequent IDFT unit. In this example an OFDM symbol is

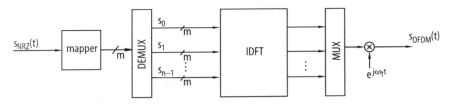

Fig. 7.27. Block diagram of an OFDM modulator

shaped from five consecutive bits. The first diagram represents the serial data
stream. After the parallel transformation each bit lies at one of the inputs of
the IDFT unit for the duration $T_U = 5\ T_B$ and generates a subsignal. The frequencies of the individual subsignals result in integral multiples of $f_0 = 1/T_U$.
They are therefore orthogonal to one another. The total of all five subsignals
results in the wave-form of the time-domain signal of an OFDM symbol after
IDFT.

$$s_{OFDM}(t) = \sum_{k=0}^{4} s_k \cdot e^{j2\pi k \frac{t}{T_U}} \cdot rect\left(\frac{t - \frac{T_U}{2}}{T_U} \right) \tag{7.50}$$

From (7.50) it can be seen that the impulse response of each subsignal transmitted is rectangular. By applying the Fourier transform the frequency spectrum of an OFDM symbol can be computed.

$$s_{OFDM}(f) = \sum_{k=0}^{4} |s_k| \cdot e^{j\varphi k} \cdot si\left(\pi T_u \left(f - \frac{k}{T_u} \right) \right) \tag{7.51}$$

In the example shown here the third bit has the value 0, so that a notch results
at the respective position in the spectrum of the output symbol. This relationship between the sequence of subsymbols before the IDFT in the transmitter
and the individual subcarriers facilitates the shaping of the transmitted signal spectrum by way of substituting predefined subsymbols.

The reason for introducing a parallel transmission of numerous subsymbols was less due to the possibility of simple spectrum shaping, than to the
fact that this solution satisfied the demand for the longest possible symbol
duration. The example given in figure 7.28 results in an extension of the symbol duration T_U by the number of the temporarily stored subsymbols. If this
figure is chosen to be very high, the number of adjacent OFDM symbols
which contribute to ISI can be reduced considerably. A total avoidance of ISI,
however, can only be accomplished by introducing a temporal guard interval
whose task is to bridge the transient effects of the transmitted signal in the receiver which are caused by the broadcasting channel. Hence, the duration of
the guard interval must be longer than the longest time delay of all echoes. It
is therefore adjusted directly to the broadcasting channel.

The effect of the guard interval can easily be explained in the time domain
[RUELBG]. The binary data stream from figure 7.28 shall serve again as input
signal for the IDFT. For convenience, we here choose its inverted form. This
means that all bits, except the third, have the value 0 and therefore do not affect the output signal. The third bit has the value 1. At the output of the IDFT
we get a purely sinusoidal signal. It is plotted in figure 7.29 as the main signal.
During the period taken up by the guard interval T_g the transmitted OFDM
symbol is periodically prolonged in a forward direction. In the two diagrams
below it there are two echo signals to be seen. They have the same shape as the

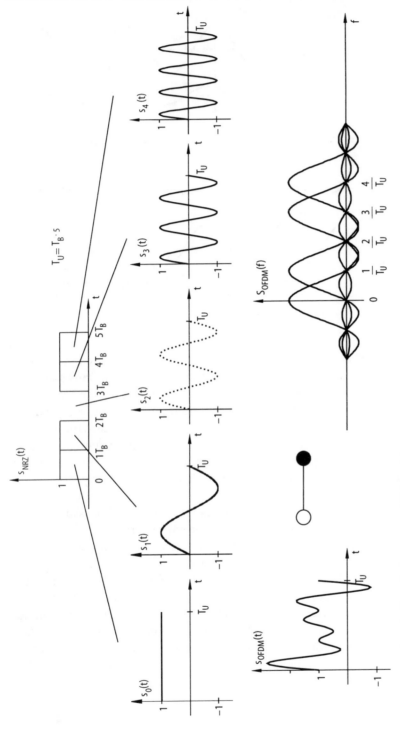

Fig. 7.28. The principle of signal processing in an OFDM modulator

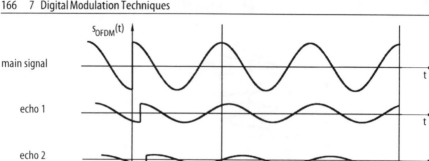

Fig. 7.29. Representation of the guard interval

main signal. Their amplitudes, however, are attenuated and their phases have been shifted due to the longer time delay. The fourth diagram shows a further signal which overlays the main signal on the transmission path. This can occur when a second transmitter sends the same signal on the same channel. The attenuation of the amplitude and the temporal difference to the main signal are relative to the distance between the transmitting stations. The co-channel signal has the same effect as a normal echo. The fifth diagram shows the addition, performed in the receiver, of all four consecutively received signals. The transients occur during the period taken up by the guard interval T_g. After all transient effects have died out, the resulting symbol assumes a stationary state which it retains throughout the entire period T_U. This period is used for the decoding of the transmitted information. In accordance with the function of the channel transmission the amplitude of the symbol has changed and its phase position differs from that of the signal transmitted. It is therefore just as important for the synchronous demodulation of the OFDM symbol to correct the effects of the channel as it is for broadband modulation techniques. The demand for an ISI-free transmission is nevertheless satisfied.

The OFDM technique is used for the transmission of DVB signals over terrestrial broadcasting channels (see chapter 11).

Symbols in Chapter 7

$A; A'$	amplitude in general
a	variable parameter for the amplitude
a_{thresh}	variable parameter for the amplitude boundary
BER_{Ix}	bit-error rate
c_{cor}	correction factor
d	Euclidean distance
$E\{x\}$	expected value of x
E_b	energy per bit
$erf(x)$	error function
$erfc(x)$	complementary error function
E_s	energy per symbol
f	frequency in general
$f(a)$	Gaussian amplitude-density distribution (normal distribution)
f_N	Nyquist frequency
f_o	basic frequency in the OFDM technique
$H(f)$	transfer function
$h_t(t)$	impulse response of the pre-filter at the transmitting end
I	in-phase component
i	running variable (integer)
j	$\sqrt{-1}$
k	running variable (integer)
l	running variable (integer)
M	amplitude states per symbol
m	bits per symbol
N	noise power
N_o	noise-power density in general
n	parameter (integer) or number of OFDM subcarriers
n_o	individual value of the parameter n
PSD_{Ix}	power spectral density
Q	quadrature component
S	mean signal power
$S(f)$	Fourier-transformed function of $s(t)$
S_a	instantaneous power
SER_{Ix}	symbol-error rate
S_{Ix}	signal power
$s_{Ix} s(t)_{Ix}$	signal in time domain
$si(x)$	$(\sin x)/x$
s_k	k subsymbol in OFDM technique
T_B	duration of one bit
T_g	duration of guard interval
T_U	duration of useful interval = $1/f_o$
T_S	duration of one symbol
t	time in general
x	variable in general
z	variable in general
α	roll-off factor
δ	δ (masking) function (Dirac impulse)
ϕ_{sk}	phase shift of the kth OFDM subsymbol
σ_R	standard deviation of a Gaussian noise signal
ω, ω_i	(i^{th}) radian frequency
ω_T	radian carrier frequency
ω_k	k^{th} radian subcarrier frequency in OFDM

8 Conditional Access for Digital Television

"Conditional access" (in the following referred to as CA) is a technique used to protect a programme or a number of programmes from unauthorised viewing. Its implementation requires a variety of technical and commercial system components, which serve the purpose of making the programmes available only to those viewers authorised to receive them (pay TV). Viewers are usually required to pay a monthly or annual fee to gain access to a particular programme channel (pay-per-channel) or, alternatively, a fee for an individual programme (pay-per-view). CA is a technique which originated, and is widely used, in English-language countries, which is why the English expressions are internationally accepted. An overview of a complete CA system is shown in figure 8.1.

The programme signal is processed in a scrambler before transmission. Within the framework of the DVB Project it has been possible to develop a so-called common scrambling system, which is supported by all CA providers. The specification describing this system is not published so that possible "pirates" will have difficulty acquiring the knowledge needed for the construction of illegal descramblers. Although the members of the DVB Project are aware that an absolutely secure scrambling system cannot be found, they are satisfied that the common scrambling system adopted is as secure as possible. As long as the instructions for deciphering are missing in a receiver, it will be impossible to view a scrambled programme.

The DVB Project has taken the initiative to propose anti-piracy laws for Europe and for each individual country. These laws will complement the development of the common scrambling system.

The concept of the common scrambling system is based on the cascading of two ciphering procedures. In the first system, data blocks of 8 bytes, each consisting of 8 bits, are scrambled, and in the second, the resulting data are re-scrambled bit by bit [ETR 289].

The procedure used for scrambling is illustrated in figure 8.2. First of all a decision is taken as to which data are to be scrambled. If, for example, scrambling is performed at the level of the transport stream the header cannot be included, because the header is necessary to synchronise the receiver. Furthermore it must be possible for the content provider to scramble only part of the services.

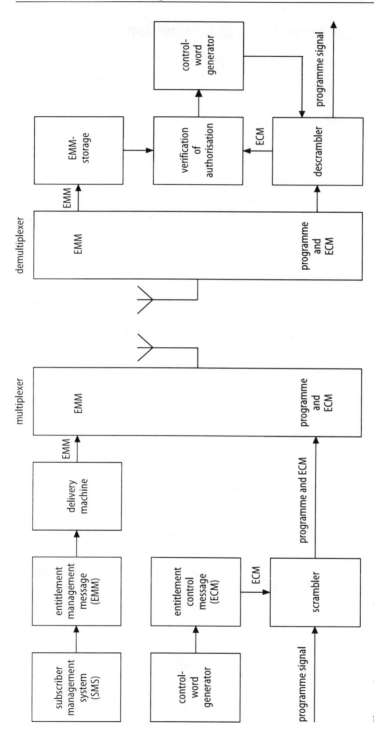

Fig. 8.1. System overview of a conditional-access system

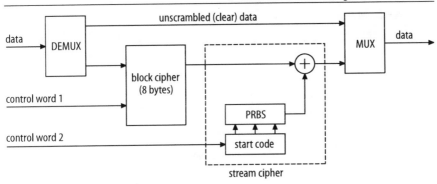

Fig. 8.2. Scrambling procedure for DVB

The first step utilises a block-cipher procedure, a technique based on 8-byte blocks. A first "control word" is required for the ciphering. The data stream coded in that way is then fed into a stream-cipher mechanism which operates with a pseudo-random generator, i.e. it creates a period of pseudo-random data out of another control word. This step can be implemented with the aid of a feedback shift register which, at a given moment, is loaded with a specified initialisation value. The bit stream which is output by this generator is then added modulo-2 to the data to be scrambled.

In the MPEG-2 structure, two levels at which the ciphering can take place are envisaged (see chapter 5): the level of the packetised elementary stream (hereafter referred to as PES) and that of the transport stream (hereafter referred to as TS). Only one of these levels should be used at any one time. The respective header, which is never ciphered, contains special control bits which have the same meaning at both levels (Table 8.1): the first shows whether the respective block is coded and the second which cipher (even or uneven) is used for the packet. This differentiation is required for the following reason. The key (i.e. the encrypted representation of the two code words) is changed from time to time. The new key is transmitted with the MPEG-2 data stream, and the second scrambling-control bit then shows that everything following the block to which the header belongs is subject to the changed cipher.

There are some restrictions for scrambling at PES level [ETR 289] with regard to the mapping of the PES onto the TS (see figure 8.3):

Table 8.1. Control bits for scrambling

Bit values	Meaning in TS and PES header respectively
00	no scrambling
01	(not used at the moment)
10	scrambled with even code word
11	scrambled with uneven code word

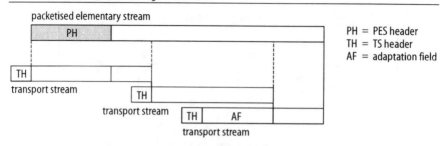

Fig. 8.3. Mapping of a PES packet onto the transport stream (TS)

- The PES header should not be larger than 184 bytes, i.e. it must fit into the useful data range of a TS packet.
- From the beginning of its header, the PES packet is divided into segments of 184 bytes each, which are then mapped onto a TS packet. These TS packets can therefore have no adaptation fields.
- If the last segment is smaller than 184 bytes, it will be preceded within the TS packet by an adaptation field of the appropriate length.
- Should an adaptation field become necessary during the transmission of a scrambled PES packet, a separate TS packet has to be inserted which contains only this adaptation field.

The aim of these limitations is to minimise the storage requirement at the receiver by simplifying the deciphering.

The two control words are required to enable the receiver to descramble the programme information. They are subjected to a separate encryption procedure at the transmission end, which transforms them into entitlement control messages (hereafter referred to as ECMs). The first ECM is used to inform the descrambler in the receiver how the blockwise scrambling can be revoked. The second ECM enables the descrambling of the bit-by-bit scrambling.

In order to transmit the information required for the descrambling in the receiver a conditional-access table has been specified as part of the so-called service information (SI) within the framework of the DVB Project. The entitlement management messages (hereafter referred to as EMMs) can be transmitted in this table as well as the ECMs. The EMMs originate from the customer administration of the CA provider, i.e. from the subscriber management system (SMS). They are attached to the programme stream via a delivery machine which, for example, ensures that each subscriber is supplied hourly or daily with the new, individually valid EMM. If the pay-per-view procedure is to be applied to a programme, then the EMM has to be made available at short notice for each programme as soon as access has been ordered by the subscriber.

The EMM is stored in the receiver. In order to enable an unambiguous allocation of the EMM to a particular receiver (i.e. to the subscriber) each receiver has to have an individual identity number, which, for example, is made available by a memory card. When the receiver has thus been authorised by EMM and memory card to receive a specific channel or programme, the control words required for the deciphering are generated from the ECMs and loaded into the descrambler, and the descrambling can commence.

The development of the common scrambling system, the use of which, as mentioned before, has been agreed within the DVB Project, constitutes a unique step towards a common user terminal for all scrambled services. This system has even been introduced into a European directive by the European Commission. All user terminals, on condition that they have the right ECMs and the correct EMM, will be in a position to descramble all scrambled programmes.

The introduction of pay programmes using common receiver hardware has thus been reduced to the question of whether the providers of scrambled programmes are willing and able to accept a further degree of standardisation. The options for further agreements are as follows:

- All providers of pay programmes agree on a uniform CA system. This option has proved to be completely unrealistic for commercial reasons.
- If every receiver were to have a common interface (CI) for an exchangeable plug-in module, or something similar, then the provider-specific processing of ECMs and EMM as well as all further steps in the CA procedure could be integrated into such a module. All providers could then work with their own CA systems without consideration of one another. This "multicrypt" possibility is generally regarded as viable. As a consequence it was possible to finalise the specification for a common interface which was standardised by CENELEC as [EN 50221]. The reason why the definition of this interface proved to be very problematic was that the knowledge of the specification, if it described an interface right inside a CA system at the receiver, would make it considerably easier for "pirates" to build illegal hardware. Moreover, the costs of such hardware might be so much reduced by the utilisation of the common interface that the commercial threshold for potential "pirates" would also be much lower. For this reason it was decided that the common interface should be placed at the signal level of the MPEG-2 transport multiplex. This entails that the exchangeable hardware must contain the whole CA system.
- If all providers of pay programmes were to agree that, although they each want to market their own receiver which is only suitable for their own CA system XY, they are prepared, for a reasonable fee, to offer every other provider the opportunity of generating their own ECMs and EMMs by means

of this CA system XY, then the objective of a uniform hardware for the subscribers could be achieved. As a result of such an agreement it would also
be possible for potential providers to decide freely when, where and for
which group of subscribers the marketing of receivers and CA services
should commence. This option, named "simulcrypt", was the subject of
considerable discussion within the DVB Project. Following lengthy media-
tion, a code of conduct has emerged which describes the contractual basis
for simulcrypt. It is expected that this code will become an integral part of
European law. The technicalities of simulcrypt are described in DVB specification [TS 101 197-1].

If, in the future, a network operator wishes to offer his customers – who may,
for example, be cable subscribers – a uniform CA concept, then, before feeding scrambled programmes into his network, he must replace the ECMs and
EMMs contained, for instance, in the satellite-transmitted pay programmes
by data of his own CA system. This type of transformation has been designated as "transcontrol". After holding yet more discussions, the members of
the DVB Project have agreed to accept the principle of transcontrol.

9 The Satellite Standard and Its Decoding Technique

Apart from the distribution of broadcast signals by terrestrial stations and cable networks, the transmission via satellite has gained considerable importance over the past few years. For DVB, satellite transmission is also of great significance. The satellite standard developed by the DVB Project was implemented by ETSI as European standard ETS 300 421 and came into force on 1-1-1995 [ETS 421, REIMERS 3].

This chapter will first discuss signal transmission via satellite in general, then look at the encoding and decoding techniques for the DVB satellite standard and, finally, introduce the most important characteristics of the system.

9.1 The Basics of Satellite Transmission

9.1.1 Transmission Distance

Satellites for the distribution of broadcast signals are located in geostationary orbit, i.e. an orbit of about 36,000 km above the equator. The orbit duration is exactly one star day, which means that, seen from the earth, the satellites appear to be stationary. Deviations from the orbit position, due to the solar wind or due to drifting caused by the uneven gravitational pull of the earth, can be corrected by the use of steering jets. The fuel supply for these steering jets is usually the component which determines the life of the satellite. Solar cells provide the energy supply for the operation of the electronic equipment. These do deliver permanent energy (during "daylight hours"), but the power is relatively low. Therefore the satellite channel, at least the downlink, i.e. the transmission from satellite to earth, must be regarded as a channel with very limited power.

The consequences resulting from this limitation will be discussed in detail at a later point.

One of the resources which is available in great abundance is the bandwidth. Communication satellites for broadcasting transmit in a frequency range between 10.7 and 12.75 GHz and in future the band from 21.4 to 22.0

GHz will possibly also be available. Satellites, therefore, have far more bandwidth available than terrestrial transmitters. Moreover, the same frequencies can be used in different orbital positions. This is due to the distinctive directional characteristic of the receiving antenna used in satellite transmission.

9.1.2 Processing on Board a Satellite

Signals which pass through a satellite are distributed to individual transponders. A transponder is the transmission channel placed between the receiving and the transmitting antennas of the satellite and consists of a number of functional units. The frequency range assigned to a transponder has a typical transmission bandwidth of 26 to 36 MHz, but the size of the bandwidth may be much larger in some satellites.

The processing of the signals within the satellite (figure 9.1) is briefly described in the following.

The uplink signals received from the earth station by the receiving antenna are first fed to a band-pass filter, which filters out the frequency range for the particular satellite. Following a pre-amplification, the signals are mixed to the downlink frequency range. There then follows a further amplification. This arrangement with two amplifiers which function at different frequency ranges prevents self-oscillation which would occur with a single amplifier [RODDY].ubsequently the broadband signal is divided into the various frequency ranges by band-pass filters, the so-called input multiplex (IMUX) filters. The signal of each of the frequency ranges is then fed into a travelling wave tube amplifier (TWTA) with a characteristic curve which increases almost linearly in the lower range, then becomes flatter and finally reaches a saturation point (figure 9.2). If the entire capacity of the amplifier tube is to be

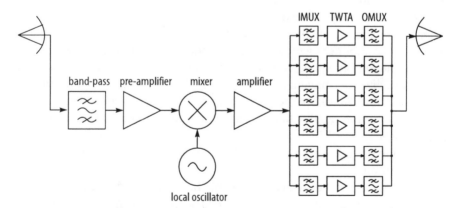

Fig. 9.1. Simplified block diagram of the payload on board a satellite

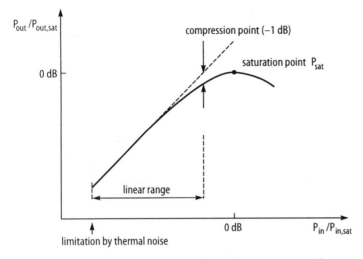

Fig. 9.2. Typical power-transfer characteristic of a travelling wave tube amplifier

used, the non-linearity has to be accepted, which has consequences for the transmission scheme, i.e. the channel coding and modulation:

– With amplitude modulation, the described non-linearity would lead to signal distortion, which means that only frequency- or phase-modulated signals are acceptable. Therefore frequency modulation, due to its immunity to interfering, is chosen for analogue television transmission, whereas for DVB a digital phase modulation (QPSK) is used (see chapter 7).

– If several signals are transmitted by one transponder in the frequency multiplex, intermodulation occurs, i.e. new frequency components are formed by the non-linear components in the amplifier. To avoid these, the operating point must be in the linear part of the TWTA's power-transfer function. This can be achieved by reducing the input power by a few dB, i.e. by applying a so-called input back-off (IBO). As a consequence of this, the output power (output back-off [OBO]) is reduced. Since the maximising of power is the key requirement, the transmitted signal needs to be one single carrier on which the payload is multiplexed in a time-division multiplex. The potential existence of intermodulation products is another reason for the necessity for each transponder to have its own amplifier and to transmit one modulated carrier only.

At the output of the power amplifier the signal from each transponder is filtered once again by an output multiplex filter (OMUX), during which process the harmonics originating in the non-linear amplifier are suppressed. The transponder signals are then recombined and fed to the transmitting antenna, downlinking it to earth.

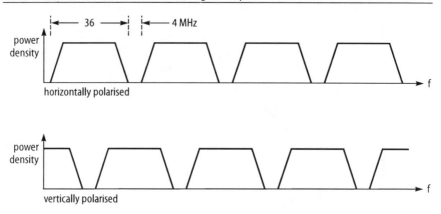

Fig. 9.3. Example of a transponder arrangement with polarisation decoupling

The transmitting antenna has a strong directional characteristic, which causes a relatively high power density in the main direction of the transmission. The power an isotropic radiator needs to have to uniformly generate a maximum of power density in all directions is called "equivalent isotropically radiated power" (EIRP) and constitutes an essential parameter for determining the characteristic features of a satellite.

9.1.3 Polarisation Decoupling

Electromagnetic waves offer the possibility of linear or circular polarisation [RODDY]:

For linear polarisation, a horizontally polarised wave and a vertically polarised wave of the same frequency may be modulated by two independent signals. At the receiver the two signals can then, ideally, be completely separated again (in practice, the crosstalk is <–18 dB). This method is used, for example, by Astra and Eutelsat. A circularly polarised wave consists of a horizontally and a vertically polarised wave of the same amplitude with a phase difference of ±90°. The sign of the phase difference determines whether the polarisation of the signal is left-handed or right-handed. Circular polarisation is used for example for TV-Sat.

Figure 9.3 shows a frequently used scheme in which the horizontally and vertically polarised transponders are staggered with respect to frequency. This has the advantage that the centre frequency of a transponder, which in analogue transmission typically has a maximum power density, faces the gaps in the other polarisation plane. Crosstalk can thus be minimised by this arrangement.

9.1.4 Energy Dispersal

In general, it cannot be assumed that the power density of a digital television signal will be distributed evenly within the transponder bandwidth. In the case of a longer sequence of ones, for instance, as might occur in stuffing packets, the power in a QPSK modulation concentrates on the carrier frequency. Such peak power in the spectrum can cause a distortion in the reception of channels of neighbouring satellites, when these transmit in the same frequency range and the directional characteristic of the receiving antenna on the earth is not narrow enough. Hence the aim is to achieve a power-density spectrum of the modulated signal that is as even as possible.

In analogue television, where these effects on the power-density concentration also occur, the energy dispersal can be achieved by an additional frequency modulation of the carrier with a triangular signal [RODDY]. To avoid interference effects in the image signal, a basic frequency of 25 Hz (50 fields per second) is chosen for PAL transmissions. The carrier oscillates continuously between two frequencies with a separation of one MHz and thereby achieves a somewhat more even energy dispersal.

In DVB the energy dispersal occurs at the very level of the code. A scrambling technique, which is described in detail further down, sees to it that the data stream assumes a seemingly random structure, which results in an almost even energy dispersal.

The energy dispersal is then revoked by the same operation being performed at the receiver.

9.1.5 Signal Reception

At the receiving end the signals transmitted from the satellite in the gigahertz range are directed via a parabolic reflector to the actual antenna (figure 9.4). They are then pre-amplified in the low-noise block (LNB) and downmixed to the first intermediate frequency (IF). This lies in the range of 950 to 2150 MHz and enables the signal to be transferred via a reasonably priced coaxial cable to the receiver. In the receiver there is an arrangement which is similar to that in the satellite: amplifier – mixer – amplifier. The mixing frequency is chosen such that the selected channel within the broadband satellite signal lies in the band pass of the following filter which suppresses all other channels (superheterodyne principle). Subsequently, the signal can be demodulated.

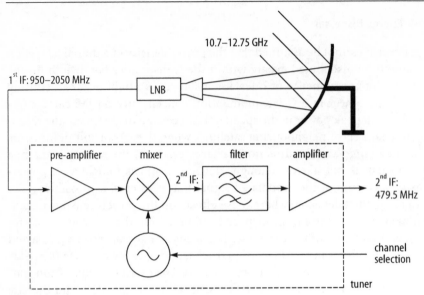

Fig. 9.4. Simplified block diagram of a satellite receiver

9.1.6 Reference Data of a Television Satellite with Astra 1D as an Example

As an example of the order of magnitude of the technical details discussed above, table 9.1 summarises the performance of satellite Astra 1D.

Table 9.1. Data of the Astra 1D satellite

Start:	October 1994
Life expectancy:	13 years
Orbit position:	19.2° east
Total power:	3300 W
Number of transponders:	18
Transponder bandwidths:	26 MHz (–3 dB) for FSS band (10.70–11.70 GHz)
	33 MHz (–3 dB) for BSS band (11.70–12.07 GHz)
Transponder output power:	63 W
EIRP:	52 dBW = 160 kW

9.2 Requirements of the Satellite Standard

The technical requirements of the transmission path on the one hand and the user demands on the other constitute the parameters for the definition of the satellite standard.

The requirements concerning the transmission path can be deduced from the following characteristics of satellite transmission:

- Due to the low-power capacity of the satellite channel the travelling wave tube should run in full saturation. Therefore modulating techniques in which the carrier is also subjected to an amplitude modulation (as for example in the higher-order QAM) cannot be considered.
- To avoid intermodulation in the travelling wave tube, only time-division multiplexing on a single carrier can be used.
- There must be energy dispersal in order to achieve a power density which is distributed as evenly as possible within the transponder bandwidth.
- The low carrier-to-noise ratio at the receiver, which is mainly due to the extremely low received power, makes a high-quality error protection necessary. A quasi error-free (QEF) transmission which is defined by a bit-error rate of less than 10^{-11} is envisaged. In practice, this corresponds to an average of about one false bit per hour per transponder.

The above requirements need to be contrasted with the demands made on the system by the users:

- For television broadcasting – as well as for other digital services – high transmission rates are demanded.
- The available transmission capacities must be used flexibly, i.e. different services with different data rates must be supported.
- The quality of the error protection, too, must be able to be adapted in a flexible way to the various requirements. For example, for a transmission via a high-powered satellite, a lower-quality error protection than that required for a satellite with a lower transmitting power will suffice, so that the net data rate (i.e. the rate of the effectively transmitted uncoded data) is higher although bandwidth and QEF requirement are the same.
- The receiving antenna should have a reflector diameter as small as possible and should be as unobtrusive as possible when mounted on the wall of a house. Furthermore, like the receiver as a whole, it should not be too expensive. However, a small receiving antenna also implies a low received power and therefore a low carrier-to-noise ratio.

Below are listed the basic characteristics of the satellite standard, which was developed taking the above requirements into account [ETS 421]:

- The source coding is carried out in accordance with the MPEG-2 standard, which combines an efficient data compression with a flexible system concept (see chapters 4 and 5).
- As already mentioned, energy dispersal at the encoding end is carried out by scrambling with a pseudo-random sequence.
- A concatenated error protection permitting various code rates is used.
- The QPSK modulation, which guarantees a constant carrier signal amplitude, was chosen for the DVB standard.

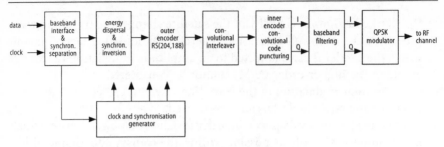

Fig. 9.5. Block diagram of signal processing at the transmitting end

- Only one carrier per transponder is used in order to avoid intermodulation.
- The combination of various services in a "data container" for simultaneous transmission via one transponder takes place in a time-division multiplex. The basis for this is the MPEG-2 transport stream.

9.3 Signal Processing at the Encoder

9.3.1 System Overview

In the following, the satellite standard will be explained on the basis of a block diagram representing the coding and modulation process at the transmitting end (figure 9.5).

The main steps in the adaptation of the transport stream to the satellite transmission link are:

- energy dispersal by scrambling,
- concatenated error protection with interleaving,
- QPSK modulation.

The following two characteristics of the transport stream are of particular significance for the coding and modulation in satellite transmission (cf. chapter 5).

- The MPEG-2 transport stream is composed of individual packets (hereafter referred to as frames) with a length of 188 bytes each. The first four bytes form the header, the first header byte being the synchronisation byte (hereafter referred to as sync byte).
- The "transport-error indicator bit" is also defined in the header. If the packet is no longer decodable due to too many channel errors, then this bit is used to indicate an undecodable erroneous packet for the source decoding.

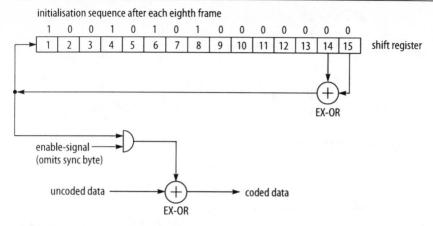

Fig. 9.6. Diagram of the energy dispersal circuit

9.3.2 Energy Dispersal

To generate a channel data stream with the most evenly distributed power-density spectrum possible (see section 9.1.4) the signals are combined bitwise with the output stream of a pseudo-random generator, via an "EXCLUSIVE-OR" operation (modulo-2 addition). This random generator is implemented by a feedback shift register, which is re-initialised at the start of every eighth frame in accordance with a predetermined bit pattern. It is only the sync byte that remains unscrambled, so as to enable synchronisation of the removal of the energy dispersal by the receiver.

When the random generator is re-initialised at the start of every eighth frame, the current sync byte is inverted as a signal for the "energy-dispersal remover" at the receiving end.

9.3.3 Error-protection Coding

The error protection coding follows the energy dispersal. This is a concatenated coding which comprises a block code, an interleaver and a convolutional coder (see chapter 6, [SWEENEY]).

The block code is an RS(204, 188) code, which enlarges each block of 188 bytes (in other words, a complete transport-stream packet) by 16 correction bytes to a gross length of 204 bytes and which can thus correct up to eight erroneous bytes.

The interleaver is designed as a convolutional interleaver with an interleaving depth of $I = 12$. The base delay is 17, and therefore the block length of the interleaver is $n = I \cdot M = 204$ bytes – in other words, a complete RS-coded

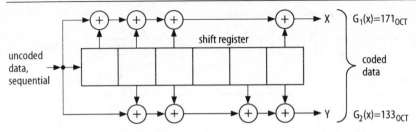

Fig. 9.7. Structure of a convolutional encoder

packet. Bytes which would originally have been contiguous in the data stream are subsequently at least 205 bytes apart.

A convolutional encoder with the basic code rate 1/2 and the generator polynomials $G_1 = 171_{OCT}$ and $G_2 = 133_{OCT}$ follows the interleaver (figure 9.7).

Several possibilities for puncturing are envisaged in order for the coding to be as flexible as possible, to adjust to the existing conditions of the channel and to the requirements of the transmission. According to the DVB specification the following code rates are possible: 1/2 (no puncturing), 2/3, 3/4, 5/6 and 7/8.

Figure 9.8 shows the arrangements which are used for each instance of puncturing. Following the puncturing of some bits from the data streams X and Y delivered by the encoder, the remaining bits must be rearranged so that the original order with two paths can be retained. These two paths are required to supply the modulator with two bits simultaneously (I- and Q-signals).

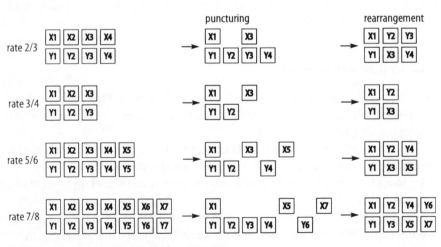

Fig. 9.8. Procedure for puncturing convolutional codes

9.3.4 Filtering

Subsequent to the error-protection coding, the data are pre-filtered and fed into the modulator.

The filtering fulfils two purposes:

- Firstly, the signals must have a band limitation in order to avoid the possibility of crosstalk on adjacent channels.
- Secondly, the filtering serves to shape the signals in accordance with the first Nyquist criterion (see section 7.1).

The Nyquist filter, introduced in section 7.1, with a raised-cosine edge and a roll-off-factor of $\alpha = 0.35$, has been chosen as the filter characteristic for the satellite standard. This filter characteristic is valid for the whole channel, which means that it is the product of all filters in the transmission path. Apart from the somewhat broader-band IMUX- and OMUX-filters in satellites this characteristic applies especially to the pre-filter in the transmitter and to the receiver input filter in the satellite tuner. In accordance with the standard, the filters in the transmitter and in the receiver must have the same transfer function, therefore this must be the square root of the total transfer function; hence these filters are called "square-root raised-cosine" filters. A template for the signal spectrum at the modulator output is given in the appendix to the satellite standard.

9.3.5 Modulation

The filtered signals of both paths are fed into a QPSK modulator [MÄUSL] as in-phase and quadrature components (figure 9.9). The so-called Gray coding allocates the bits to the respective constellation points. If, at demodulation –

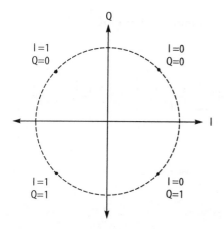

Fig. 9.9. Constellation diagram of the QPSK

owing, for example, to noise in the channel – a phase is decoded which passes beyond one of the two decision thresholds (= axis in figure 9.9), there will only be one erroneous bit.

Before transmission, the IF signal is upconverted to radio frequency.

In this way two bits per symbol are transmitted; hence in ideal circumstances the utilisation of the bandwidth, referred to the gross bit rate, is $B = 2$ bit/(s · Hz). However, owing to the non-ideal filters (finite width of the filter slopes) the bandwidth utilisation is actually smaller. Depending on the configuration of the system, the ratio between bandwidth and symbol rate (BW/R_S) would be in the range of about

$$\frac{BW}{R_s} = 1.27 \, \frac{\mathrm{Hz}}{\mathrm{symbols}/\mathrm{s}}. \tag{9.1}$$

Hence the bandwidth utilisation would be about

$$B = \frac{\mathrm{bits\ per\ symbol}}{BW/R_s} = \frac{2\,\mathrm{bits}/\mathrm{symbol}}{1.27\,\mathrm{Hz}/\frac{\mathrm{symbols}}{s}} = 1.57 \, \frac{\mathrm{bit}}{\mathrm{s\ Hz}}. \tag{9.2}$$

The bandwidth utilisation can also be referred to the net bit rate, i.e. to the number of information bits only, without the error protection. In this case the calculated value, as indicated above, must be multiplied with the code rates of the outer and inner codes.

9.4 Decoding Technique

On reception of the signals transmitted from the satellite, the processing steps that were carried out at the encoder have to be reversed. In particular, the errors which occurred in the channel must be corrected. Moreover, it is necessary to recover the synchronising information, which is required for the channel decoding.

The decoder consists of the following components (figure 9.10):

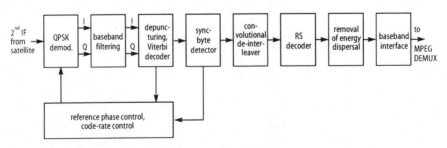

Fig. 9.10. Block diagram of signal processing at the receiving end

- demodulator,
- filter,
- clock recovery,
- sync-byte detector,
- decoder and interleaver for concatenated error protection,
- energy-dispersal removal,
- baseband interface,
- control unit.

9.4.1 Demodulator

The QPSK demodulator must first of all recover the carrier frequency from the input signal. A particular problem is the phase uncertainty of the carrier, which must be resolved in order to demodulate I and Q correctly. The input signal can assume one of four possible phase positions, each of which represents two bits (see figure 9.9) and can serve as a reference for the detection, so that it is possible to have four carrier phases at right angles to each other. As the absolute phase position is unknown to the receiver, it first chooses a phase at random for use in the demodulation.

The demodulator can be implemented together with the carrier recovery by means of a Costas loop [MÄUSL, COSTAS]. The carrier frequency for demodulation is created by a voltage-controlled oscillator (VCO). The VCO is operated by a control signal which is generated, in a control loop, from the two demodulated baseband signals.

The decision as to whether the right phase position is in use can only be made later, in two steps, in the decoder. A phase error of ±90° can be resolved in a first step, and in a second step the residual uncertainty of 180° can be removed. The mechanisms required will be described later in the appropriate sections.

The correction of the absolute carrier phase recognised as wrong generally does not happen at the recovered carrier itself, because this would require considerable expenditure. A correction of 90° can be obtained by exchanging the I and Q components after demodulation and subsequently inverting one of them. A phase error of 180° is compensated by an inversion of the bit stream at the position in the decoder at which it was identified.

9.4.2 Filtering and Clock Recovery

Following demodulation, the baseband signals are subjected to the square-root raised-cosine filtering, as was the case at the transmitting end (see section 9.3.4). Thereafter crosstalk between adjacent symbols at the sampling point is minimal.

The filter can be implemented as a digital filter after an oversampling of the demodulated baseband signal or as an analogue band-pass filter before the demodulator.

The clock of the filtered baseband signal can be recovered with the aid of a PLL-circuit so that the signal can be sampled [LEE].

9.4.3 Viterbi Decoder

The first step in error-correction is the decoding of the convolutional code, performed by the Viterbi decoder, which is preceded by the depuncturing (see section 6.3.4). Soft decision must be made available to the Viterbi decoder by the demodulator. Soft decision improves the error-correction capability and makes possible the correct interpretation of depunctured bits. In most cases 3 bits are implemented, which, in the case of a code rate of 1/2, results in a reduction of the required signal-to-noise ratio by approx. 2 dB for a residual bit-error rate of $2 \cdot 10^{-4}$ (figure 9.11, cf. section 9.5.2). A resolution higher than 3 bits is hardly an improvement and only increases the implementation requirement (see figure 9.11). However, 4-bit soft-decision quantisation can be found in state-of-the-art implementations.

During the processing, the Viterbi decoder counts the number of errors that have been identified and transmits this number to the control unit which has the task of determining the as yet unknown transmission parameters:

– the reference phase for the demodulation (possible uncertainty of 90°),
– the puncturing scheme,
– the synchronisation of the depuncturing. Similar to the puncturing, the depuncturing is a block-oriented periodic process (see figure 9.8). Com-

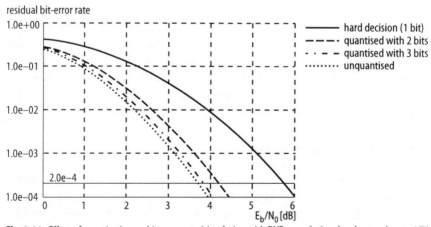

Fig. 9.11. Effect of quantisation on bit-error rate (simulation with DVB convolutional code at code rate 1/2)

mencing at the wrong point, i.e. not at the beginning of a block, will result in the depunctured bits being also inserted at the wrong points.

The determination of these – as yet unknown – parameters is achieved by means of the control unit working through all possible combinations. At the output of the Viterbi decoder the data are then recoded and punctured, in accordance with the parameters determined at the decoding, and subsequently compared with those bits which were originally available at the depuncturing input. In this way the value of the bit-error rate can be computed. Should only one of the above-mentioned parameters be wrong, then this rate would be quite high, due to the fact that the redundancy generated in the encoder would be wrongly interpreted.

It is only when all the parameters are correct that a proper decoding is possible and that the measured bit-error rate will assume a value which is considerably smaller and roughly corresponds to the actual bit-error rate of the channel. The precondition for this is that the signal-to-noise ratio be high enough, so that there is a sufficient distinction between a correct and a false choice of parameters. If the satellite link operates at least close to the standard operating point defined by DVB, the procedure will lead to the desired result. However, if the signal-to-noise ratio is not sufficient, so that the bit-error probability is too high even if the parameters are correct, the decoder will carry on searching. If all possibilities have been exhausted a corresponding signal will be supplied which informs the user that reception is not possible.

A special case in the detection of the missing parameter is the shifting of the reference phase in the demodulator by 180° as against the reference phase in the transmitter. This will result in a bit stream at the receiving end which is inverted compared with the bit stream that was transmitted. The Viterbi decoder will then also emit a bit stream, which (apart from uncorrected errors) is the inverse of the uncoded bit stream from the transmitter. The phase error cannot be detected, hence the uncertainty of 180° remains and is only resolved in the subsequent module, which is the sync-byte detector.

9.4.4 Sync-byte Detector

Further decoding requires that the subdivision of the data stream into the MPEG-2 TS packets as well as the 8-packet structure for the energy dispersal be detected. The sync byte at the beginning of each frame is required for this. At the transmitting end the sync byte remains unchanged by RS encoder, interleaver and energy dispersal and is therefore available at the output of the Viterbi decoder.

The task of the sync-byte detector is to detect the regular occurrence of the sync byte, or of its inversion, every 1632 bits (204 bytes) within the bit stream and from this to derive synchronising signals for the de-interleaver, the RS decoder and the energy-dispersal remover. In the course of this operation the last phase of uncertainty can now be removed from the decoding process: the occurrence of seven inverted sync bytes and one non-inverted sync byte indicates an erroneous phase shift of 180°, in which case the bit stream is simply inverted at the output of the detector.

A byte clock is also derived so that, after a serial-to-parallel conversion, further processing can take place at the byte level.

9.4.5 De-interleaver and RS Decoder

The satellite channel is essentially a Gaussian channel, which means that the signal has been corrupted by additive white Gaussian noise, and that bit errors occur singly rather than in bursts. The convolutional code is well-suited for the correction of such errors. However, in the case of an accumulation of bit errors the Viterbi decoder will fail and supply a wrongly decoded signal, i.e. it will in its turn generate a burst error. To correct this burst error, interleaving and outer error protection (Reed-Solomon code) are applied. If as a result of a decreasing signal-to-noise ratio the number of bit errors in the transmission increases, the bursts become longer and more frequent. Finally, if more than eight erroneous bytes occur in one frame at the input of the RS decoder, the error protection will fail. In most of these cases the decoder recognises that the data can no longer be corrected and transfers an appropriate signal to the subsequent circuits.

Uncorrected errors generally become apparent by distorted blocks appearing in the picture. When the signal-to-noise ratio reaches the critical value at which even the outer coder starts to fail, the number of errors increases very fast in the event of a further drop in the signal-to-noise ratio, until at last even the synchronisation fails and a decoding of the picture is no longer possible. This total breakdown occurs in the range of a signal-to-noise ratio of less than one dB.

9.4.6 Energy-dispersal Remover

Energy dispersal takes place at the transmitting end by means of the modulo-2 addition of a pseudo-random number sequence (see section 9.3.2) and is reversed in the receiver by the same operation. It follows from this that the circuit in the receiver is the same as that in the transmitter. To initiate the pseudo-random generator the energy-dispersal remover requires a signal which indicates the start of the 8-packet sequence and is supplied by the sync-byte detector.

9.4.7 Baseband Interface

After the original MPEG-2 TS has been reconstructed it can be transferred to the TS demultiplexer which feeds the various components of the data stream into the respective source decoders. For the signalling of errors which have occurred in the transmission channel and have been identified by the error protection but which the error protection was unable to correct, MPEG-2 has provided a specific signal: the "transport-error indicator bit", which is the first bit after the sync byte and will be set in such cases. This bit enables the demultiplexer to identify the respective packets as erroneous and take the necessary precautions.

9.5 Performance Characteristics of the Standard

9.5.1 Useful Bit Rates

The useful bit rate R_u can be computed from the symbol rate by

$$R_u = R_S \frac{\text{bit}}{\text{symbol}} R_1 R_2 \tag{9.3}$$

where $R_1 = 188/204 = 0.922$ is the code rate of the Reed-Solomon code and R_2 is the code rate of the convolutional code with or without puncturing.

Table 9.2 [REIMERS 3] lists the possible useful bit rates for a system with $BW/R_S = 1.27$ as a function of the transponder bandwidth and the code rate.

Several members of the Astra and Eutelsat satellite families have commenced transmission of DVB signals, for example via 36 MHz transponders with $BW/R_S = 1.309$ and with a code rate R_2 of 3/4. This results in a symbol rate of 27.5 Mbaud and a useful bit rate of 38.01 Mbit/s.

Table 9.2. Useful bit rate [Mbit/s] in the satellite channel

Channel bandwidth [MHz]	Symbol rate [Mbaud]	Maximum useful bit rate [Mbit/s] at a varying code rate R_2				
		1/2	2/3	3/4	5/6	7/8
54	42.5	39.2	52.2	58.8	65.3	68.5
46	36.2	33.4	44.5	50.0	55.6	58.4
40	31.5	29.0	38.7	43.5	48.4	50.8
36	28.3	26.1	34.8	39.1	43.5	45.6
33	26.0	24.0	31.9	35.9	39.9	41.9
30	23.6	21.7	29.0	32.6	36.2	38.1
27	21.3	19.6	26.2	29.4	32.7	34.4
26	20.5	18.9	25.2	28.3	31.5	33.1

Table 9.3. Internal code rate and required carrier-to-noise ratio

Code rate R_2	Required E_b/N_o	Corresponding C/N (incl. 0.8 dB)
	[dB]	[dB]
1/2	4.5	4.1
2/3	5.0	5.9
3/4	5.5	6.9
5/6	6.0	7.9
7/8	6.4	8.5

9.5.2 Required Carrier-to-noise Ratio in the Transmission Channel

As already explained in section 9.2, it is the aim of the error protection to achieve a bit-error rate of $\leq 10^{-11}$ subsequent to the decoding. For this a bit-error rate of $\leq 2 \cdot 10^{-4}$ is required at the output of the Viterbi decoder. This in turn depends on the code rate and the carrier-to-noise ratio at the receiver input. Table 9.3 [REIMERS 3] shows the corresponding values for the satellite standard which result from a simulation, based on the assumption of an additional loss of 0.8 dB through the practical implementation margin (of modulator, demodulator and TWTA). This takes into account the fact that the hardware components used in practice cannot perform as postulated in theory, that for example the sampling points in the demodulator are slightly changed by the phase noise of the recovered carrier.

The following holds for the above values (see chapter 7):

E_b/N_o = energy per information bit/noise-power density
C/N = carrier-to-noise ratio.

The "information bits" are defined as the uncoded bits (i.e. before error-protection processing) which are set in relation to the overall energy of the coded bit stream. In this way the fact is taken into account that only part of the transmitted data (all requiring transmission power) are useful data.

With QPSK and concatenated coding at the rates R_1 and R_2, the relation between E_b/N_o and C/N is

$$C/N = E_b/N_o \, 2R_1 R_2 \tag{9.4}$$

where factor 2 takes the number of 2 bits into account which are transmitted within one symbol of the QPSK.

The value C/N is dependent on the following conditions for both the uplink and the downlink:

- transmitting power,
- precision of the alignment of the transmission and reception antennas,

Table 9.4. Required antenna diameter for some selected system configurations

Channel bandwidth [MHz]	Data rate [Mbit/s]	Antenna diameter in cm for an average-year service continuity		
		99.7%	99.9%	99.99%
	Code rate 2/3			
54	52.2	50	58	88
33	31.9	39	45	69
26	25.2	35	40	61
	Code rate 5/6			
54	65.3	63	72	111
22	39,9	49	56	86
26	31.5	44	50	77

- diameter of the reception antennas,
- meteorological conditions,
- noise figure of receivers.

9.5.3 Antenna Diameter

Examples of the required antenna diameters are shown in table 9.4 [REIMERS 3]. The data are valid for the hydrometeorological zone E [CCIR 563, CCIR 564] in which Germany is located. Other zones have other parameters due to varying climatic conditions. Further, a downlink frequency of 12 GHz and an EIRP of 51 dBW have been assumed. The service continuity refers to 99.7%, 99.9% and 99.99% respectively, for an average year. The last figure, for example, stands for an average loss of 53 minutes in one year during which time the receiving conditions, for instance due to a thunderstorm, are so poor that the error protection fails completely.

Due to the fact that in the case of a drop in the carrier-to-noise ratio the image quality remains at first unchanged (see section 9.4.5), the breakdown comes as a surprise and could be wrongly interpreted by the viewer as a device error, especially since many or all channels might be affected at the same time, depending on the error protection. For such cases special mechanisms should be installed in the equipment to warn the viewer when the carrier-to-noise ratio drops below the critical level.

9.6 Local Terrestrial Transmission

The satellite standard introduced here is not only conceived for use in satellite transmission, but also for the transmission of MMDS in the frequency band above 10 GHz.

MMDS is a microwave multichannel/multipoint distribution system and operates as a local terrestrial transmission. A microwave transmitter at an elevated point, such as a church spire, covers a relatively small area within a radius of less than 50 km. Reception is possible with a small fixed antenna in line-of-sight with the transmitter. For this the frequency bands envisaged are between 2 and 42.5 GHz, but these differ from country to country.

For transmissions below 10 GHz what matters most is a full utilisation of the available bandwidth, therefore the DVB cable standard (see chapter 10) is used for the transmission (DVB Microwave cable-based – DVB-MC [EN 749]). Above 10 GHz the air causes the attenuation to increase considerably with increasing frequency [RODDY], which calls for a higher robustness of the transmission, as provided by the DVB satellite standard (DVB Microwave satellite-based – DVB-MS [EN 748]).

The fact that the two MMDS standards are based on the DVB-S and DVB-C standards, respectively, ensures that the hardware developed for the latter two can be reused for demodulation and decoding; only the antenna and the tuner need to be adapted.

Symbols in Chapter 9

b_i	bit stream no. i
B	bandwidth utilisation
BW	bandwidth
C/N	carrier-to-noise ratio
E_b	energy per useful bit
f	frequency in general
$G_i(x)$	i^{th} generator polynomial of a convolutional code
I	interleaving depth or in-phase component
i	running variable, integer
IF	intermediate frequency
M	interleaver delay
n	integral control variable, code-word length of a block code
N_o	noise-power density in general
P_{out}	output power
P_{in}	input power
P_{sat}	saturation power
Q	quadrature component
R_i	i^{th} code rate
R_s	symbol rate
R_u	useful bit rate
t	time in general
X	1st output branch of a convolutional encoder
x	argument for a generator polynomial
$X_1, X_2, \ldots$	n^{th} bit of the 1st output branch of a convolutional encoder
Y	2nd output branch of a convolutional encoder
$Y_1, Y_2, \ldots$	n^{th} bit of the 2nd output branch of a convolutional encoder
α	roll-off factor

10 The Cable Standard and Its Decoding Technique

The specification for the transmission of DVB signals in cable networks was developed in the DVB Project between August 1993 and January 1994 [STEN-GER]. The result was the draft of a baseline system. In the spring of 1994 this was recommended by ETSI as a European transmission standard and in November 1994 it became an ETSI standard, designated as ETS 300 429 [ETS 429]. The specification was also submitted to the International Telecommunication Union (ITU). In June 1995, ITU recommended that it be used as a standard which should not only be valid for Europe.

10.1 Cable Transmission Based on the Example of the German Telecom CATV Network

In Germany the cable-based transmission of broadcasting signals from the studio to the user equipment is divided into four sections by the national reference chain. A simplified structure of this chain is shown in figure 10.1.

The supraregional section connects the studios to each other or feeds the signals from a studio to a TV switching station of the public network. In the regional section the signals are distributed to the feeding points of the CATV networks in the cable head-ends. The local section 1 represents the CATV distribution network. The subsequent in-home network and private cable networks are referred to as local section 2.

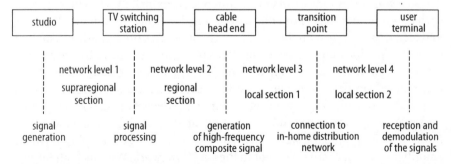

Fig. 10.1. Simplified structure of the reference chain in Germany

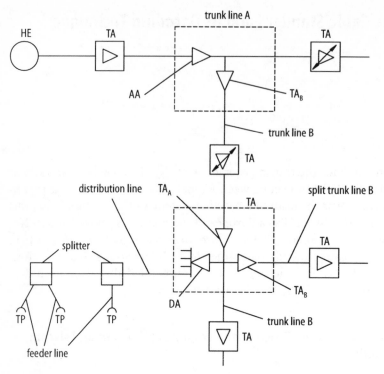

Fig. 10.2. Schematic representation of the CATV network

The CATV network constitutes a tree-and-branch system which is divided into the 4 network levels A, B, C and D [KENTER]. One branch of this structure can be seen in figure 10.2. Network level A starts with its trunk line just after the cable head-end. After a maximum cable length of 412 m an amplifier point (TA) is installed, which basically consists of a cable equaliser and a trunk amplifier. Up to 20 trunk amplifier points can be connected in series at network level A and are all interconnected by a cable route. For the trunk lines of network level A this results in an upper limit of approximately 8 km in length. At each amplifier point it is possible to decouple the signal from network level A and feed it to B (or also directly to C), in which process the signal is attenuated by 14 dB. The first amplifier after the decoupling has the task of compensating for the 14 dB of the decoupling attenuation. The adjacent network level B is constructed similarly to network level A. Each third trunk amplifier point is amplitude-controlled by a pilot signal with a frequency of 80.15 MHz. By this means a temperature-dependent drifting of the amplification is avoided. From network level B the signal is fitted to the C level via a C amplifier. This C amplifier is the last active element in the CATV network. With its high output level it ensures that the signal, subsequent to passing through the passive C

Table 10.1. Index of abbreviations used

Abbreviation	Designation
HE	head-end
TA	trunk amplifier point
TA_A	trunk amplifier of network level A
TA_B	trunk amplifier of network level B
DA	distribution amplifier
TP	transition point

level and also the passive D level, has a guaranteed level at the transition point (TP).

The requirements for the transmission quality of television and radio broadcast FM signals are stated in [FTZ 1] and [FTZ 2] for each section in the national reference chain. No official specifications of requirements for digital signals have so far been established. Therefore we will estimate in the following sections the effect of the overlay of intermodulation and random noise signals on the digital signal.

10.1.1 Intermodulation

In general, transmissions via CATV networks are limited by intermodulation products. In the case of non-linear processing these are caused mainly by the interaction of the image carriers of the analogue TV signals transmitted in different channels. Depending on the frequency position or the spectral distance of the disturbance from the image carrier, the resulting patterns in the picture can produce very coarse structures, which moreover move randomly. These disturbances can easily be perceived by the human eye. This is the reason that the reference chain ensures a signal-to-interference ratio of 60 dB at the user outlets.

However, the digital signal is less sensitive to narrow-band interference. The symbols received are corrupted by an intermodulation signal only if the highest amplitudes of the disturbances exceed the decision thresholds of the symbol decision performed at the receiving end. The power ratio between useful signals and unwanted signals in this case is of course dependent on the type of modulation used. If the peak amplitude of the sinusoidal signal is scaled to value 1, then, according to (7.42), the following expression results if, for instance, a QAM signal is used as the useful signal:

$$\frac{C}{I} = \frac{S_{QAM}}{S_{sin}} = \frac{2\frac{M-1}{3}}{\frac{1}{2}} = \frac{4}{3}(M-1). \tag{10.1}$$

Hence the following holds for a 64-QAM with $M = 64$ symbol states:

$$10 \log\left(\frac{C}{I}\right)\Bigg|_{64\text{-}QAM} = 10 \log(84) = 19.24 \text{ dB}. \tag{10.2}$$

If there were no further unwanted signals in the cable network apart from the intermodulation interferers, the signal-to-interference ratio C/I would be allowed to be below 20 dB in the case of a 64-QAM. This also implies that, if the values guaranteed by the FTZ guideline [FTZ 1] were duly observed, the level of the digital signal could be reduced by approximately 40 dB as compared to the level specified for the analogue television signal. Therefore intermodulation disturbances will have no effect on the deterioration of the digital signals.

However, it is to be noted that additional signals fed into the existing networks present an increased burden to the cable amplifiers. Since analogue services will be transmitted over cable networks for many years yet and since the picture quality of these services must not noticeably deteriorate, the total power of the high-frequency composite signal, particularly in the intensely used networks, must not be unduly increased by the introduction of DVB signals. Field tests showed that this condition is satisfied if the DVB signals are fed into networks at a reduced level, which is 10 to 13 dB below the level specified for analogue television signals [SCHAAF].

10.1.2 Thermal Noise

A further deterioration of the useful signal is caused by the superposition of noise. The reference chain defines a signal-to-noise ratio of 44.5 dB at the receiver input for the analogue video signal. The noise figure of this ratio relates to the effective values of a baseband signal which is band-limited at 5 MHz and shaped by a noise-weighting filter. The effect of various factors must be taken into account when converting this data into a carrier-to-noise ratio C/N [VELDERS], during which the correction of the signal power and the noise power can be dealt with separately.

The following characteristics are important for the signal: the carrier attenuation at the point of 50% amplitude response in the filter at the receiving end, the level control of the black-and-white range with 70% of the CVBS signal amplitude, a 90% depth of modulation, and the conversion from peak value to effective value.

The power of the noise signal is changed by the IF filter at the receiving end and by the noise-weighting filter.

The multiplication of the two correction factors results in a factor of approx. −1.5 dB:

$$10 \log\left(\frac{C}{N}\right)\Bigg|_{5\,MHz} = 44.5 \text{ dB} - 1.5 \text{ dB} = 43 \text{ dB}. \tag{10.3}$$

Let the signal-to-noise ratio of the digital signal apply to an 8-MHz channel. Taking into account the aforementioned 13-dB difference value of the level, the result is a power ratio between DVB signal and noise of

$$10 \log\left(\frac{C}{N}\right) = 10 \log\left(\frac{C}{N}\right)\Bigg|_{5\,\text{MHz}} -10 \log\left(\frac{8\,\text{MHz}}{5\,\text{MHz}}\right) -13\,\text{dB} \cong 28\,\text{dB}. \qquad (10.4)$$

Hence, at the introduction of DVB services, this carrier-frequent signal-to-noise ratio is the least that can be expected in a broadband cable network.

For the estimation of the quality of digital signals one generally uses the bit-error rate of the information bits. This can be computed as a function of the ratio E_b/N_0 as outlined in chapter 7, so that now a connection will be established between the C/N and E_b/N_0 supplied by the transmission channel.

It is the matched filter in the receiver which has a most decisive effect on the connection between these two ratios. The conditions set by (7.27) and (7.37) on the one hand, and the identity of the signal power in transmission channel C and in baseband S on the other hand, result in the following equation:

$$\frac{C}{N} = \frac{E_b}{N_0} \cdot m. \qquad (10.5)$$

As the noise bandwidth of the matched filter in the receiver is twice the Nyquist frequency $2\,f_N$, this results in a reduction of the effective noise power subsequent to passing through the filter. On condition that the spectrum of the transmission channel is optimally utilised this value can also be expressed by the roll-off factor α:

$$\frac{2f_N}{B} = \frac{1}{1+\alpha}. \qquad (10.6)$$

When digital information is transmitted in real systems the modulation is always supported by an error-correction algorithm (see chapter 6). The previously introduced redundancy R is, however, removed again from the signal in the receiver, so that the effective signal power is reduced accordingly. This results in the desired relation

$$\frac{C}{N} = \frac{E_b}{N_0} m \cdot R \frac{2f_N}{B}. \qquad (10.7)$$

The following set of parameters which are used in the DVB standard (see section 10.3) shall serve as an example:

$m = 6$ for a 64-QAM,
$R = 188/204$ for a Reed-Solomon error protection with the code rate R,
$2f_N/B = 1/1.15$ for a roll-off factor of 0.15 (see (10.6)).

The CATV network ensures an E_b/N_0 of approx. 21 dB if these parameters are chosen.

10.1.3 Reflections

A cable connection between two network components (e.g. two splitters in a passive network) is not generally ideally terminated. The signal travels along the cable path, for instance from a first splitter to a second. It can be conceived as a forward-travelling wave [UNGER]. At the connection point between the cable and the second splitter a portion of the signal power is reflected and returns along the cable path. Hence this portion is also called backward-travelling wave. The amplitude ratio between the go- and the return wave can be indicated by the return loss. For the components used in a CATV network a frequency-dependent quantity of 15–20 dB is required. When travelling through the cable path the reflected signal undergoes a second attenuation. Given that, as a rule, even the output of the first splitter is not totally identical with the characteristic impedance, a portion of the reflected signal power is reflected in its turn. The doubly reflected signal portion has a cumulative effect on the forward-travelling wave since it moves in the same direction as the main signal. It is delayed by the amount of time it requires for travelling twice through the cable path. Of course a portion of its power is again reflected at the connection point between the cable and the second splitter. The attenuation that an echo undergoes at both connection points and while travelling along the cable is generally so high that reflections of a higher order, i.e. signal portions which have travelled forward and backward more than once, can be ignored. The result is a transfer function of the following form:

$$H(f) = \frac{1 + \varrho \cdot e^{-j2\pi f_c \tau} \cdot e^{-j2\pi\Theta}}{\sqrt{1 + \varrho^2}} \tag{10.8}$$

where

ϱ	amplitude of the echo,
f_c	carrier frequency,
τ	delay between echo and main signal,
Θ	additional phase shift of the echo signal.

The return loss of at least 15 dB of the components used, on the one hand, and an assumed low cable attenuation of only 2 dB, on the other, results in an amplitude of the echo of –34 dB, relative to the main-signal amplitude. Investigations [DIGISMATV] have shown that with a 64-QAM technique reflections can be neglected if the relative amplitude is smaller than –30 dB.

10.2 User Requirements of the Cable Standard

When the CATV network of Deutsche Telekom (German Telecom) was built up in the 1970s the basic supply of television signals to the public was by means of terrestrial transmission. The design of the network parameters for CATV was therefore based upon the parameters required for terrestrial systems. The chronological development of the various standards within the DVB Project began with the drawing up of the satellite specification [ETS 421]. The development of the cable specification [ETS 429] did not begin until after the first results of the satellite transmission emerged. There was a catalogue of user requirements which had to be taken into account:

(1) The CATV network, in its conventional configuration, must be usable for digital signals.
(2) The feeding of the DVB signals into the cable networks must not noticeably impair the quality of the usual analogue services.
(3) As many data as possible should be transmitted on one cable channel so as to ensure an adequate useful data rate which is compatible with all satellite channels.
(4) Cost is a crucial parameter in the introduction of a new system. This applies not only to content providers but also to network operators. Most important, of course, are the costs to be borne by the consumer. For this reason a high priority in the development of systems was the reasonable pricing of decoder hardware in the integrated receiver/decoder (IRD). The cable IRD, just as the IRD for terrestrial reception (see chapter 11), should be as similar as possible to the satellite IRD.
(5) It should be possible for the cable IRDs to be introduced on the market at the same time as the satellite IRDs.

The conditions for the future transmission of DVB signals were established by requirements 1 and 2 above. The adaptation of the signal to the existing cable channel is effected by a QAM technique which yields the best results in terms of systems theory (see chapter 7).

A study of the available network structures concerning master antenna television systems revealed the coexistence of various network configurations and quality classes. As the programmes distributed in these systems are mainly received from a satellite or satellites, the systems are called satellite master antenna television (SMATV). In accordance with the requirement expressed in point 5 it was decided to develop an additional standard [ETS 473] which extended the CATV standard so as to make it applicable to SMATV systems.

The modelling of the cable standard on the satellite standard, as required under points 3 and 4, is achieved by the input signal processing. From baseband interface via energy dispersal and use of the outer error protection to convolution interleaver, the signal processing is identical in both systems. For the cable standard the application of an internal error protection is not required. There are two main reasons for this. Firstly, each additional process

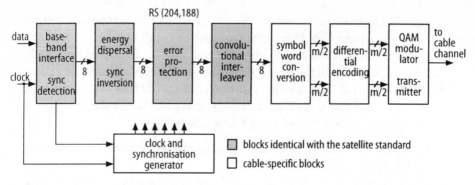

Fig. 10.3. Block diagram of a DVB cable encoder

that inserts redundancies into the data stream reduces the useful data rate, which is contrary to requirement 3. Secondly, the cable networks are of a high quality (see section 10.1), so that the DVB signal is sufficiently protected against transmission errors by the use of the outer error protection.

10.3 Signal Processing at the Encoder

This section introduces the DVB cable specification as defined in the ETSI standard [ETS 429]. As only the signal processing in the encoder is described in this standard, the present section will be restricted to the discussion of the processing blocks at the encoder. Moreover, the signal technique, from baseband interface to convolution interleaver, will not be reiterated here, as these were already discussed in the chapter dedicated to the satellite standard. The DVB cable encoder is shown as a block diagram in figure 10.3. The shaded processing blocks are identical with those in the satellite standard (see section 9.3).

The cable networks, contrary to the situation in the case of satellites, ensure a relatively constant and high transmission quality. For this reason the cable standard provides for several bits to be combined to form a symbol and to be transmitted by a higher-level quadrature amplitude modulation. There is a choice between five variants. The modulation type with the lowest efficiency provided in the standard, i.e. an efficiency of 4 bits per symbol, is 16-QAM, followed by 32-QAM which combines 5 bits per symbol. The modulation type with the highest efficiency in the first version of the standard was the 64-QAM, which enabled a parallel transmission of 6 bits per symbol. The first revision of the standard provided for a transmission efficiency of 7 and 8 bits per symbol with a 128-QAM and a 256-QAM respectively.

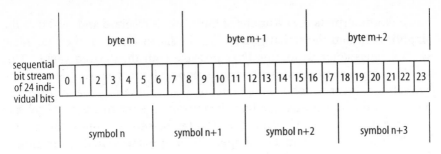

Fig. 10.4. Conversion of the bytes to symbol words, taking the 64-QAM as an example

10.3.1 Conversion of Bytes to Symbol Words

After the 8-bit data words have passed through the channel coding they are fed into a symbol-word converter (referred to as "byte-to-m-tuple converter" in the standard). This has the task of combining the m bits which are to be transmitted per symbol into symbol words of a corresponding width. During transmission with 16-QAM the individual bytes are divided into two 4-bit symbol words. The MPEG transport packets, which are augmented from 188 bytes to 204 bytes in the Reed-Solomon encoder by appending 16 redundancy bytes (see section 6.4.3), have a length of 408 symbol words after conversion of the symbol words. For the conversion of the 8-bit data words to 6-bit symbol words, as required when using 64-QAM, the individual bits must be reordered beyond the byte boundaries. Four symbol words result from 3 data words, so that 24 bits form one framework. The packet length following the reordering increases to 272 symbol words. Figure 10.4 shows the conversion of a 24-bit framework, using the example of a 64-QAM.

In the case of the 32-QAM the fixed frame length is augmented to 40 bits. After the reordering of the 5-bit symbol words the 204 bytes of an extended transport packet result in a frame length of 326.4. This means that a 204-byte packet cannot be assigned to a whole-number frame length.

10.3.2 Differential Coding of MSBs

Following the conversion of the 8-bit data words to symbol words, the latter are differentially pre-processed as described below. The reason for this measure is to be found in the loss of absolute phase information by the suppression of the carrier signal at the transmitter. The receiver can only recover the reference phase up to an uncertainty of $n \cdot 90°$, due to the known signal statistics of the QAM technique (see figure 7.22). In order to achieve this, a frequency-synchronous demodulator separates the received signals into two components which bear an orthogonal relation to each other. Following this, the demodulator interprets the positions of the symbols received and readjusts the amplification and the signal phase until the mean values of the individual symbol states coincide as closely as possible with the desired values. This leads to a constellation as described in section 7.5, figure 7.24. The values of the two subsignals correspond to the values of the in-phase and quadrature components of the transmitted symbol. The information relating to the algebraic sign of both components is, however, lost by the suppression of the carrier. For this reason the information is specially pre-processed in the transmitter, firstly, by separating the two MSBs from the rest of the actual symbol word and, secondly, by differential coding of only these MSBs, which are given in (10.9) by A_K and B_K:

$$I_K = (\overline{(A_K \oplus B_K)} \cdot (A_K \oplus I_{K-1})) + ((A_K \oplus B_K) \cdot (A_K \oplus Q_{K-1}))$$
$$Q_K = (\overline{(A_K \oplus B_K)} \cdot (B_K \oplus Q_{K-1})) + ((A_K \oplus B_K) \cdot (B_K \oplus I_{K-1})).$$

(10.9)

I_K and Q_K are referred to as the results of the Boolean operation, i.e. the differentially encoded MSBs, and their temporal predecessors are I_{K-1} and Q_{K-1}. As according to (10.9) the outputs of the processing block are fed back into the inputs, this is a recursive algorithm. The truth table 10.2 illustrates the dependence of the output variables on the input variables.

If, for example, the values $A_K = 0$ and $B_K = 0$ are valid at the input of the differential encoding unit, at the output there result the same values I_K and Q_K as the ones of the symbol previously transmitted. As both values have remained unchanged, the actual symbol is transmitted in the same quadrant as the one temporally preceding.

According to table 10.2, if the input occupancy of $A_K = 0$ and $B_K = 1$, I_K assumes the value which was allocated to the inverted Q-component in the previous symbol, while Q_K assumes the value which the I-component previously had. This results in a 90° phase shift in a mathematically positive direction. The constellation diagram known from the chapter on satellite transmission (figure 9.9) shall serve as an example, as its allocation requirements concerning bits and quadrants also apply to the cable standard. If I_{K-1} is defined as having had the value of 0, and Q_{K-1} the value of 1, then the previously

Table 10.2. Truth table for differential encoding

Inputs		Outputs		Rotation
A_K	B_K	I_K	Q_K	
0	0	I_{K-1}	Q_{K-1}	0°
0	1	$\overline{Q_{K-1}}$	I_{K-1}	+90°
1	0	Q_{K-1}	$\overline{I_{K-1}}$	−90°
1	1	$\overline{I_{K-1}}$	$\overline{Q_{K-1}}$	180°

transmitted symbol had just been in the 4th quadrant. In the case under consideration, i.e. $A_K = 0$ and $B_K = 1$, both I_K and Q_K have the value 0 (see table 10.2). The actual symbol can be found in the first quadrant in accordance with figure 9.9. The phase shift from the centre of the fourth quadrant to that of the first quadrant is identical with the +90° rotation as indicated.

The advantage resulting for the receiver from the differential pre-processing is that, for recovering the MSBs, the receiver only needs to evaluate the changes in those quadrants in which a transmitted symbol was/is found. The receiver carries out a subtraction, in which process the absolute phase information is dropped. Hence the phase uncertainty of $n \cdot 90°$ as described above suffices for the recovering of the MSBs. For the sake of completeness it should be mentioned that the LSBs, which have not yet been discussed, possess the information about the state of the symbol within the quadrant selected by the MSBs. Figure 7.22 shows the constellation of a 64-QAM. In the first quadrant the 16 possibilities are drawn as vectors resulting from the four remaining LSBs.

10.3.3 Modulation

The feeding of the symbol words with the differentially encoded MSBs into the quadrature amplitude modulator is the final step in the processing chain of the encoder. The multiplication with a complex carrier signal results in a band-pass signal, at the output of the modulator, whose centre frequency is equal to that of the carrier signal. This frequency is referred to as intermediate frequency (IF). Concerning the modulation, it is particularly important that the unequivocal allocation of the symbol words to the complex amplitudes of the symbols to be transmitted be guaranteed. Depending on the number m of the bits which are to be transmitted per symbol, a constellation results in which a definite amplitude and phase position are modulated onto the carrier by every possible bit combination (see chapter 7). Figure 10.5 shows the allocation for the 16-QAM, the 32-QAM and the 64-QAM as defined in the standard. The 16 amplitude phase states of the 16-QAM have been plotted as a proper subset directly in the constellation diagram of the 64-QAM. In the 6-bit symbol words of the 64-QAM it is necessary to delete only the two middle bits b_4 and b_3 in order to achieve an allocation of the 4-bit words to the amplitude phase conditions of the 16-QAM. Here again it can be clearly seen that the MSBs of the individual bit combinations never change within one quadrant and therefore are of no relevance to any decision taken within a quadrant.

The bit combinations of the symbol words for the 16-QAM and the 64-QAM have been chosen such that if a symbol state is altered in a way that causes it to take on a neighbouring value only one single bit is changed. This,

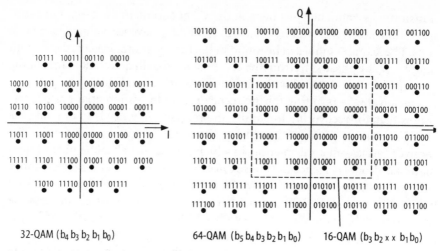

Fig. 10.5. Constellation of the 16, 32 and 64-QAM

however, only holds for those cases in which the invalidation causes no crossing of quadrant borders. Within one and the same quadrant the symbol words are thus coded in accordance with Gray [LÜKE 1]. If the transmitted symbol has been invalidated so as to assume a state which lies in a neighbouring quadrant, there can occur up to 5 bit errors per symbol in the DVB signal. Compared to a system that is entirely Gray-coded this results in a disadvantage, which, however, cannot be avoided because of the introduction of the differential coding. The decoding algorithm requires a rotationally symmetrical arrangement of the individual quadrants; however, within the quadrants it enables the implementation of a Gray code.

The revised version of the standard includes a 128-QAM and a 256-QAM, the constellation diagram of the latter being shown in figure 10.6.

The layout of the matched filter at the receiver is described in section 7.1. The roll-off factor defined in the cable standard has the value

$$\alpha = 0.15 \ . \tag{10.10}$$

The transfer function shown in figure 7.6a refers to the total transmission path. Hence, in the pre-processing at the transmitting end there must be a subfilter with a square-root raised-cosine characteristic.

There is a template in the annexe to the ETSI standard which gives the required frequency response of the amplitude of all filters used in the transmitter (see figure 10.7). Apart from the matched filter, the effect of the D/A conversion and the transfer functions of all further filters need to be taken into account. The resulting in-band ripple in the pass band may not exceed 0.4 dB, while an out-of-band rejection greater than 43 dB is required in the stop band.

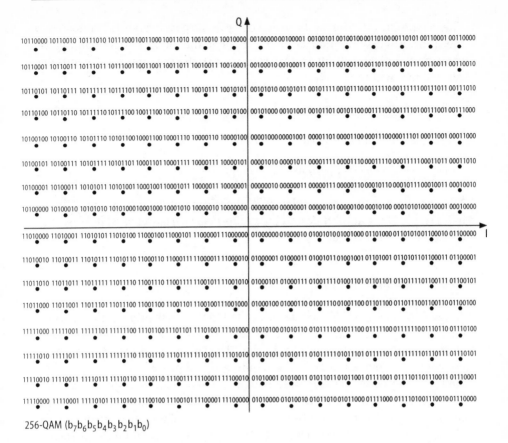

256-QAM $(b_7 b_6 b_5 b_4 b_3 b_2 b_1 b_0)$

Fig. 10.6. Constellation of the 256-QAM

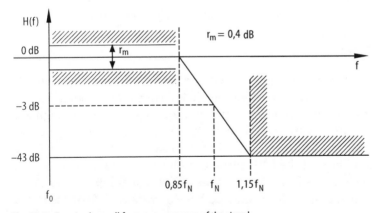

Fig. 10.7. Required overall frequency response of the signal

The maximum variation of the group delay in the pass band is 10% referred to the transmitted symbol duration T_S. In a channel of an 8-MHz bandwidth, utilising the given roll-off factor from (10.10), the maximum transmittable symbol rate will be

$$R_S = 2f_N = \frac{B}{1+\alpha} = 6.96 \text{ Mbaud} \qquad (10.11)$$

which results in a symbol duration T_S of

$$T_S = \frac{1}{2f_N} = \frac{1+\alpha}{B} = 143.75 \text{ ns}. \qquad (10.12)$$

The variation of the group delay shall therefore not exceed 14.4 ns. The in-band ripple of the amplitude and the group-delay variation must be observed up to the Nyquist frequency f_N.

10.4 Decoding Technique

As the ETSI standard ETS 300 429 is limited to the description of the processing algorithm in the encoder there are various possibilities for the realisation of the individual system units in the decoder. These sometimes differ fundamentally from one another, as the required signal processing can be carried out in the digital as well as in the analogue domain. The input interface of the cable tuner and the cable tuner itself are basically similar to the structural components used in conventional receivers for analogue television signals. The input signal is converted from its RF range to a constant intermediate frequency in accordance with the superheterodyne principle. At this point the channels are separated with the aid of a steep-edged intermediate filter. For the subsequent quadrature amplitude demodulation the suppressed carrier signal must be recovered. Furthermore, the sampling of the transmitted signal requires a recovering of the clock signal used at the transmitter.

10.4.1 Cable Tuner

The cable tuner converts the signals received, after a pre-filtering by means of a tunable selection filter, to a fixed intermediate frequency (IF) range. Here it is of no importance whether an analogue television signal or a DVB signal is concerned. For cost-saving reasons it seems sensible to use the same frequency converter for the conversion of both kinds of signals. However, it has to be borne in mind that the DVB signal cannot only be transmitted with a bandwidth which differs from the commonly used channel spacing of 7 or 8 MHz, but in principle also with a bandwidth of, for instance, 2 MHz (see table 10.5).

In order to achieve what is generally known in analogue television as the IF characteristic [SCHÖNFD 2] a VSB-modulated signal from the RF transmission channel must be down-converted with a sinusoidal signal of a local oscillator whose frequency is 38.9 MHz above the image-carrier frequency. Given that the upper as well as the lower sideband of the DVB signal need to be transmitted, the suppressed carrier in this case is placed at the centre of the transmission channel. Hence a difference of 2.75 MHz in frequency results between the two carrier positions. If for both the analogue and the digital signal the same oscillator is to be used for the conversion to the IF range, then the difference in frequency between the image carrier of the analogue TV signal and the imaginary DVB carrier remains unchanged, so that, under the given conditions, the following DVB IF ensues:

$$f_{IF} = 38.9 \text{ MHz} - 2.75 \text{ MHz} = 36.15 \text{ MHz}. \tag{10.13}$$

The cable tuners used in conventional receivers are usually only adjustable to frequencies whose value is an integral multiple of 62.5 MHz. That is why the DVB IF will usually be 36 MHz.

Concerning this method it has to be borne in mind that the signal spectrum is inverted after the frequency conversion. This phenomenon is already known from analogue television. All portions of a signal which in the HF range had frequencies above the carrier frequency, lie below the carrier frequency in the IF range and vice versa. Mathematically these changes can be described as the inversion of the quadrature component. Thus the constellation of the complex-modulated QAM symbols (see figures 10.5 and 10.6) is mirrored on its I-axis. In order to achieve an error-free decoding the necessity of this new correspondence between the complex amplitudes of the received symbols and their respective symbol words has to be taken into account. However, if the transmitter also carries out a spectrum inversion, as is common in analogue television, then the signal spectrum will be inverted twice and will therefore be back in its original state.

10.4.2 IF Interface

The actual channel selection is carried out in the IF range, as already mentioned. The conventional filters used in this range are constructed as surface acoustic-wave (SAW) filters. Owing to the required VSB Nyquist edge and the suppression of the modulated audio signals, the IF filter for the analogue TV signal differs from the IF filter for the DVB signal, and therefore it is indispensable, at this point in the process, to separate the two signal paths. The IF filter for the DVB receiver can have half the Nyquist frequency response. Since it may be implemented as a band-pass filter its transfer function needs to run symmetrically to the carrier frequency position. Due to the fixed distance between the two points of 50% amplitude response, only a signal whose symbol rate is exactly twice the Nyquist bandwidth can, as shown in (10.11), be optimally filtered. The same is true if after demodulation an analogue Nyquist filtering is carried out in the baseband. It is only through the use of a digital filter that a flexible adaptation to various symbol rates is possible. In this case the IF filter only serves to separate adjacent channels and should have a steep-edged transfer function.

10.4.3 Recovery of the Carrier Signal

A synchronous demodulation process requires a coherent carrier. This must be regenerated by a local oscillator in the receiver and synchronised with the signal received. A possible situation would be a frequency- and phase-selective control loop, on the principle of the phase-locked loop (PLL), working on a predetermined IF level. The carrier signals which are orthogonal to each other in phase are generated by means of an oscillator. According to whether the PLL is to be used for analogue or digital signals it would either be a voltage-controlled oscillator (VCO) or a digitally-controlled oscillator (DCO). The information about the phase difference between the carrier which was used in the transmitter for the modulation and the carrier which was generated in the receiver is obtained by a phase discriminator. This can be realised as an analogue or a digital unit, exactly like the loop filter.

An example of an analogue technique is shown in figure 10.8a. Non-linear processing must ensure that a discrete line is originated in the spectrum of the QAM signal at the carrier frequency or an integral multiple of the carrier frequency. However, a quadratic characteristic curve will not suffice, as

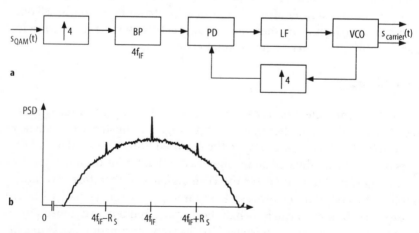

Fig. 10.8. Example of a technique for carrier recovery by quadrupling the IF range. **a** Block diagram, **b** Power spectral density in the range of the quadruple IF

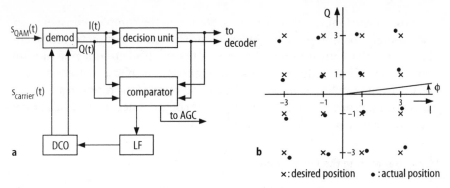

Fig. 10.9. Example of a technique for carrier recovery by decision feedback of the symbol error.
a Block diagram, **b** Effects of phase error in the constellation of a 16-QAM

shown by the PSD in figure 10.10b. For this reason the signal is affected by a non-linear characteristic curve of the fourth order which may consist of a series connection of two squarers. Besides the various sum and difference frequencies this process yields the desired line at the quadruple carrier frequency. This range of the PSD is shown in the diagram of figure 10.8b. The signal is then preselected by a band-pass filter and serves the subsequent PLL as a reference variable input. With its quadruple frequency it has a phase uncertainty of $n \cdot 90°$, which, however, is of no importance because of the differential coding of the DVB signal (see section 10.3.2). The controlled variable is generated by a quadrupling of the frequency of the output signal of the oscillator. After a phase comparison in the phase discriminator (PD) an error signal results at the output. Subsequent to passing through the loop filter this error signal is used to lock the oscillator.

The recovery of the phase difference between the carrier signal suppressed in the transmitter and the carrier signal generated in the receiver can alternatively be performed directly in the baseband by means of the evaluation of the complex signal amplitudes (see figure 10.9). For this the QAM signal is demodulated with two carrier signals which are orthogonal in phase. At first these have a random reference phase position. After sampling with double the Nyquist frequency each symbol is compared with the M known reference values in the decision unit and assigned to the reference value which has the minimal Euclidean distance d. These reference values are then made available to the decoder for further processing. Within the loop a comparator follows the decision unit, which computes the sum and phase of the minimal Euclidean distance d. Besides the value for the phase error a value for the erroneous amplitude results. This value can be fed to the automatic gain control (AGC) as a reference signal. The phase error signal passes through a loop filter and then controls the DCO. Using the example of the 16-QAM, figure 10.9b shows how a phase error causes the received symbols – indicated here by dots as actual positions – to undergo a constant rotation in the complex constellation. Each desired position is shown by an x. For reasons of symmetry this technique has a phase uncertainty of $n \cdot 90°$. If the received QAM signal has not been converted correctly to the desired frequency, the phase deviation accumulates from sample to sample and the constellation of the received signal rotates around its own centre. This error can be compensated by the PLL as long as the frequency offset is within the lock-in range.

10.4.4 Generating the Clock Signal

Apart from the carrier signal, the clock signal with which the digital symbols in the transmitter are processed must also be generated in the receiver. The information about the clock is present, for example, in the envelope of the modulated QAM signal and can be recovered by non-linear signal processing. Therefore similar laws apply as for the recovering of the reference carrier (see section 10.4.3). The basic block diagram of a technique which relies on purely analogue signal processing is shown in figure 10.10a. A simple squarer in front of the phase discriminator PD is sufficient as a non-linear pre-processing block, as compared with that required for carrier recovery. The PSD of the signal at the output of the phase discriminator is depicted in diagram b of figure 10.10. The spectral ranges of the sum and difference frequencies which were generated by the multiplication of the IF signal with itself are easy to identify. The envelope of the QAM signal additionally generates a discrete line in the spectrum at the frequency which corresponds with the symbol rate transmitted. This line is identical with the clock frequency used at the trans-

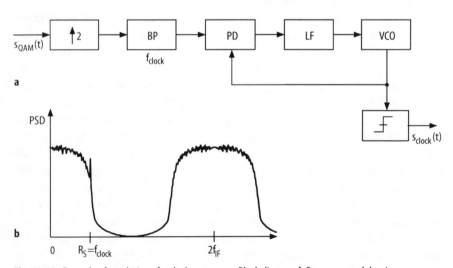

Fig. 10.10. Example of a technique for clock recovery. **a** Block diagram, **b** Power spectral density

mitting end. The synchronisation of the clock frequency generated at the receiver is effected by means of a band-pass filter and a subsequent PLL, where the centre frequency of the oscillator corresponds with the clock frequency, which in turn corresponds with the symbol rate. After an amplitude limitation the clock signal is ready for the digital signal processing.

There are further techniques described in the literature [LEE, PROAKIS, KAMMEYER, MÄUSL] which generate the error signal for the oscillator within the analogue as well as within the digital range.

10.4.5 Demodulation of the QAM Signal

The demodulator has the task of converting the QAM signal to the baseband and simultaneously splitting it into its in-phase and quadrature components. After the temporal multiplication with two carrier signals which are orthogonal to each other in phase, the resulting sum-frequency signals of the two components must be suppressed by one low-pass filter each. If the multiplication and the filtering are achieved by means of analogue signal processing, then both components will subsequently be available for A/D conversion in the baseband.

By using conventional analogue technology a reasonably priced demodulator can be manufactured. The disadvantage of this solution is to be found in the two parallel paths through which the in-phase and quadrature components have to pass before sampling as well as in the parallel sampling itself.

The parallel signal paths can be transferred to the digital level if the modulated signal is fed into a single A/D converter in its entirety. In this case the demodulation is carried out by the multiplication with two digital carrier signals. The required sampling rate for the sampling of the QAM signal at the IF level, in accordance with the sampling theorem, will have to be at least twice that of the highest frequency to be found in the signal. For transmission in an 8-MHz channel the resulting minimal sampling frequency would be

$$f_{min} = 2\left(f_{IF} + \frac{B}{2} \right) = 2\left(36.15 \text{ MHz} + \frac{8 \text{ MHz}}{2} \right) = 80.3 \text{ MHz}. \tag{10.14}$$

The high clock rate has various disadvantages when used in consumer receivers (e.g. high cost, high power consumption), so that a totally digital solution must be ruled out at this moment.

A demodulation alternative arises if the QAM signal from the IF range at 36.15 MHz is converted to a second IF range. The centre frequency of the signal should be chosen so as to correspond to the transmitted symbol rate. This technique offers the advantage that the converted QAM signal as a whole can be oversampled by a clock signal of a frequency which is exactly four times the value of the transmitted symbol rate. In the case of a symbol

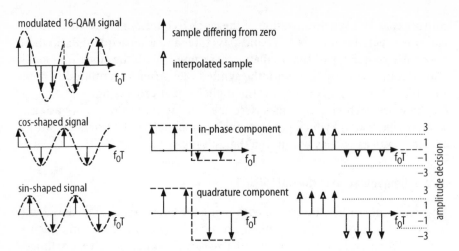

Fig. 10.11. Time graph of signals during processing in a system with a lockage between the suppressed carrier signal and the transmitted symbol rate with subsequent quadruple oversampling

rate of 6.875 Mbaud, as mentioned in section 10.5.1, this results in a clock frequency of

$$f_O = \frac{1}{T_O} = 4\,R_S = 4 \cdot 6.875 \text{ MHz} = 27.5 \text{ MHz}. \tag{10.15}$$

Hence there are exactly four samples for each modulated symbol. Figure 10.11 shows idealised signal forms. The first graph shows the received QAM signal after oversampling, the vertical arrows indicating the oversampled amplitude values. For better understanding it also shows the shape of the equivalent analogue signal. Two symbols of a 16-QAM signal are used as an example. The digital carriers required for the demodulation can be seen in the second and third graphs. These also have four samples per period: 1, 0, –1, 0 for the cosine-wave oscillation, and 0, 1, 0, –1 for the sine-wave oscillation, respectively. Here, too, the cosine waves and sine waves of the equivalent analogue signals have been plotted. The in-phase and quadrature components are generated by multiplying the sampled QAM signal with one of the two digital carriers. These can be seen in the middle diagram of figure 10.11. The same result is achieved by replacing the multipliers with a simple inverter which inverts the QAM signal for the second half of each symbol period. A subsequent demultiplexer divides the signal into two partial data streams in which every second sample equals zero.

It follows that an individual symbol, viewed in isolation from the data stream, has a mean value not equal to zero and an AC component whose direct wave has a period duration of half the length of the symbol duration. The fundamental frequency is therefore identical with the Nyquist frequency. The

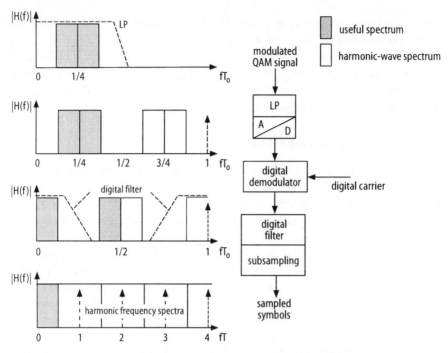

Fig. 10.12. Frequency spectra of signals processed in a demodulator if the sampling rate has the quadruple value of the symbol rate

digital representation of the signals also results in the corresponding harmonic-wave spectra.

The signal processing described is shown again in the frequency domain in figure 10.12. The signal spectra here are rectangular templates. The signal spectrum shown in the second diagram is generated subsequent to sampling with the oversampling clock. The centre frequency of the QAM signal in the useful spectrum is $fT_O = \frac{1}{4}$. Apart from this the image frequencies of the first harmonic-wave spectrum are plotted at $fT_O = \frac{3}{4}$. Following the multiplication with the digital carrier signals the useful spectrum shifts to the baseband. At the same time a second signal portion is formed at the frequency $fT_O = \frac{1}{2}$ (see third diagram). This portion of the spectrum, which consists of the sum of the carrier frequency and the frequencies of the QAM signal, causes every second sample in the wave-form of the in-phase and the quadrature component to be equal to zero. Each second sample must therefore be suppressed by a subsequent digital low-pass filter. In the time domain, the filtering results in an interpolation of the missing samples. Utilisation of one of the four samples suffices for the subsequent amplitude decision. Therefore the signal is subsampled by a factor of four. Since the oversampling was higher by a factor of four than that required according to Nyquist, this subsampling just satisfies the first Nyquist criterion (see section 7.1).

In reality no rectangular-shaped wave-forms occur, so that the instants of the subsampling – or rather, Nyquist-sampling – have to be carefully chosen. Finally, the conversion to the transmitted bit combination, the demapping of the symbol, is carried out with the help of a table.

10.4.6 Differential Decoding of MSBs

The transmitted m-bit symbol words must be differentially decoded in the receiver, i.e. the differential encoding of the symbol words (see section 10.3.2) performed in the transmitter must be revoked in the receiver. For this, as a first step, the two most significant bits of each m-bit symbol word transmitted are separated from the rest of the symbol word. The actual values I_K and Q_K have resulted in the encoder by the combination of their temporal predecessors I_{K-1} and Q_{K-1} with the actual MSBs of the uncoded symbol words A_K and B_K. These must now be recovered in the decoder from the transmitted I_K and Q_K. The rule for computing this process is found by solving the system of equations specified in the standard (see (10.9)) for A_K and B_K:

$$A_K = (\overline{(Q_K \oplus I_{K-1})} \cdot (I_K \oplus Q_{K-1})) + ((I_K \oplus I_{K-1}) \cdot (Q_K \oplus Q_{K-1})) \qquad (10.16)$$

$$B_K = (\overline{(I_K \oplus Q_{K-1})} \cdot (Q_K \oplus I_{K-1})) + ((I_K \oplus I_{K-1}) \cdot (Q_K \oplus Q_{K-1})).$$

In accordance with (10.16) the values received for I_K and Q_K are to be stored for one clock period and are thus denoted as I_{K-1} and Q_{K-1}. By a comparison with those values of I_K and Q_K which are received with a one-symbol delay, information can be gathered as to whether the successively transmitted symbols are located in the same quadrant or in different quadrants. The actual values for A_K and B_K result from the rotation of the symbols.

10.4.7 Conversion of Symbol Words to Bytes

Before the information received can be fed to the error-correction unit it has to be reconverted to its original form of 8-bit data words. Care must be taken that the same data-packet structure is generated as existed in the transmitter before the symbol-word conversion (see section 10.3.1). In the case of the 64-QAM, for example, there are three different possibilities of performing the required allocation (see section 10.4). Information is required by which the original byte boundaries are recognised. This information can be gained from the MPEG sync bytes. Therefore each IRD requires a sync-byte detector which searches the data stream for the required bit combination – 47_{HEX} for the sync byte, or $B8_{HEX}$ for the inverted sync-byte – and transfers the information to the symbol-word converter at the receiving end.

10.4.8 Detection of MPEG Sync Bytes

The detection of the MPEG sync bytes is carried out before the error correction, hence in the uncorrected data stream. At this point a typical maximum bit-error rate (BER) would be $2 \cdot 10^{-4}$. A sync-byte detector will search the incoming data stream for data words with the values 47_{HEX} or $B8_{HEX}$. In principle, two kinds of false decisions can occur. Firstly, there is a certain probability that the transmitted sync words will be invalidated by the superposition of unwanted signals in the channel. Assuming that, on average, the disturbances affect all bits to the same extent, then, by approximation, the following sync-byte error rate holds:

$$P_{e,Sync} = 1-(1-BER)^8 \xrightarrow[BER=2 \cdot 10^{-4}]{} 1.6 \cdot 10^{-3}. \tag{10.17}$$

Secondly, there are data words within the transport packets which have the above-mentioned sync-byte values as useful information. If all data words have a pseudo-random character and if all values occur with the same regularity, then an occurrence probability for a data byte of the value 47_{HEX} or $B8_{HEX}$ can be established as follows:

$$P_{47} = P_{B8} = 0.5^8 = \frac{1}{256} = 3.91 \cdot 10^{-3}. \tag{10.18}$$

While the first type of error prevents the detection of a transmitted packet, the second type wrongly indicates to the detector the beginning of a packet.

The length of the packets before passing through the error correction is exactly 204 bytes. Apart from the sync byte it comprises 187 useful bytes and 16 redundant bytes. Assuming that each one of the 203 useful and redundant bytes can take on one of the two sync-word values independently of each other, this will result in an occurrence probability of at least one value of 47_{HEX} or $B8_{HEX}$ within one packet:

$$P_{e,Packet} = 1-(1-P_{47})^{203} \cdot (1-P_{B8})^{203} \cong 0.8. \tag{10.19}$$

This implies that, through the second type of error alone, approx. 8 out of 10 packets can be erroneously detected.

By exploiting the temporal redundancy in the periodic packet structure of the MPEG data this unacceptable number can be considerably reduced. The detector output is then only activated if a random combination of the two values 47_{HEX} or $B8_{HEX}$ occurs exactly n times during a packet length. The probability of all 203 combinations occurring (as a function of n) is computed as:

$$P_{e,Packet} = 1-(1-P_{47}^n)^{203} \cdot (1-P_{B8}^n)^{203}. \tag{10.20}$$

The probabilities for $n = 2, 3$ and 4 are given in table 10.3.

Table 10.3. Occurrence probability of data words
of the value 47_{HEX} or $B8_{HEX}$ as a function of the number
of MPEG packets required for the detection

n	2	3	4
$P_{e,Packet}$	$6.18 \cdot 10^{-3}$	$2.42 \cdot 10^{-5}$	$9.32 \cdot 10^{-8}$

The first type of error, however, leads to less favourable effects. The probability that with a non-ideal channel at least one erroneous sync byte will occur among the n consecutive ones increases with increasing n.

$$P_{e,Sync} = 1 - (1 - BER)^{8n} \cong n \cdot 1.6 \cdot 10^{-03} . \tag{10.21}$$

When dimensioning a detector, which might be realised as a correlator, both error probabilities must be taken into account as a function of the number n. The whole of the following decoder processing relies on safe sync information, which has already been described in sections 9.4.5 and 9.4.6 using the example of the satellite decoder.

10.5 Performance Details of the Standard

When using the DVB standard for the cable-based transmission of DVB signals there are two aspects of particular importance. The first is to know how much useful information can be transmitted, practically error-free, over a transmission channel with a given bandwidth, and the second is to know the amount of the required carrier-to-noise ratio.

10.5.1 Determination of Useful Data Rates

In accordance with (10.11) the maximum symbol rate which can be transmitted in an 8-MHz channel is limited to a value of 6.96 Mbaud by the roll-off factor $\alpha = 0.15$ which is specified in the standard. The gross bit rate results from multiplying the number of bits m transmitted per symbol. The gross bit rates for various QAM techniques are given in table 10.4. To obtain the useful data rate, the portion of the data rate provided for the error correction must first be deducted. The DVB standard uses a Reed-Solomon error protection of code rate $R = {}^{188}\!/\!_{204}$. Since the data rate can only be given in combination with a channel bandwidth, it is usual to divide the useful data rate by the channel bandwidth. The resulting value is referred to as spectral efficiency. Its unit is bit/s per Hz. The spectral efficiency, which is only theoretically possible, is identical with the number of bits transmitted per symbol. Through the introduction of the error correction the values decrease by the

Table 10.4. Performance details of various QAM techniques

QAM technique	Number of bits per symbol m	Gross bit rate in one 8-MHz channel [Mbit/s]	Useful bit rate in one 8-MHz channel [Mbit/s]	Bandwidth efficiency [bit/s per Hz]
256-QAM	8	55.65	51.28	6.4
64-QAM	6	41.73	38.45	4.8
32-QAM	5	34.78	32.05	4
16-QAM	4	27.82	25.63	3.2

factor R, and through the utilisation of a matched filter they decrease by the factor $1/(1 + \alpha)$.

The original plan of Deutsche Telekom was for the signals to be fed into the CATV network at a symbol rate of 6.875 Mbaud [SCHAAF]. A transmission with 64-QAM would have resulted in a gross data rate of 41.25 Mbit/s and, after deducting the error protection, a useful bit rate of 38.01 Mbit/s. This is exactly the bit rate transmitted in each transponder in DVB satellite transmissions (see section 9.5.1). Signals which are contributed to a head-end as described above can generally be fed into the cable network in two ways. One option is to feed a complete data stream in its entirety into a cable channel after its having come from a transponder, subsequent to a QPSK demodulation, a Viterbi decoding, and a QAM remodulation. In this case it is not necessary to carry out de-interleaving, RS-decoding, energy dispersal, and the respective processing steps at the encoder. The other option is to combine new MPEG transport streams from various partial data streams, which in their turn were transmitted in different transponders, and to subsequently feed them into various cable channels. In this second case a complete decoding is necessary. The simultaneous utilisation of at least two transponders must be possible. The signals transmitted in the transponders are not generally in synchronism with each other. The exact values of their actual data rates are subject to the usual tolerances. As the multiplex units operate on a synchronous time base, the individual MPEG transport streams have to be adapted to the highest transmitted data rate. This can be achieved by inserting stuffing packets into the individual data streams until the data rates of the various signals are identical. For this reason Deutsche Telekom has increased the symbol rate for a DVB transmission by cable from 6.875 Mbaud to 6.9 Mbaud. This results in a sufficient useful data rate of approximately 38.15 Mbit/s. The corresponding gross data rate is 41.4 Mbit/s.

Table 10.5 is taken from appendix B of the standard and gives further examples of a system configuration. It illustrates the flexibility of the DVB cable standard. Because there is a possibility of combining the channel bandwidth

Table 10.5. Examples of system configurations

Useful bit rate [Mbit/s]	Gross bit rate [Mbit/s]	Symbol rate [Mbaud]	Bandwidth used [MHz]	Modulation scheme
38.1	41.34	6.89	7.92	64 QAM
31.9	34.61	6.92	7.96	32 QAM
25.2	27.34	6.84	7.86	16 QAM
31.672 PDH	34.367	6.87	7.90	32 QAM
18.9	20.52	3.42	3.93	64 QAM
16.0	17.40	3.48	4.00	32 QAM
12.8	13.92	3.48	4.00	16 QAM
9.6	10.44	1.74	2.00	64 QAM
8.0	8.70	1.74	2.00	32 QAM
6.4	6.96	1.74	2.00	16 QAM

and the number of bits transmitted per symbol, the useful data rate can be increased from approx. 6 Mbit/s to more than 38 Mbit/s. With the DVB cable standard it is possible to achieve a data rate of, for instance, 31.672 Mbit/s by using at least a 32-QAM. Therefore the standard is also compatible with the terrestrial PDH (plesiochronous digital hierarchy) networks.

10.5.2 Carrier-to-noise Ratio Required in the Transmission Channel

The spectral efficiencies of the modulation types specified in the standard can be seen in figure 10.13 (except for the 128-QAM), each as a function of the required carrier-to-noise ratio. A bit-error rate of $2 \cdot 10^{-4}$ was chosen for this example. In this case the power of the subsequent error correction is just sufficient for a practically error-free signal to be interfaced at the decoder output (see section 6.2.6). The theoretical values for a redundancy-free transmission in accordance with section 7.5 are marked in figure 10.13 as crosses. The conversion of E_b/N_0 to C/N was carried out in accordance with (10.5). The positions of the DVB signals are represented by dots, for which purpose (10.7) was used to compute the C/N values. Each transfer from the theoretical signal to the DVB signal improves the carrier-to-noise ratio by approx. 1 dB. These improvements are mainly due to the introduction of the matched filter in the receiver, which results in a decrease of the actually effective noise power (see Section 10.1). This gain, however, is paid for by a reduction in the useful data rate, which manifests itself in a lower bandwidth efficiency.

Furthermore, figure 10.13 shows the upper limit of possible C/N values, which follows from (10.4) if the DVB signals are fed into the cable networks with a power reduction of 13 dB.

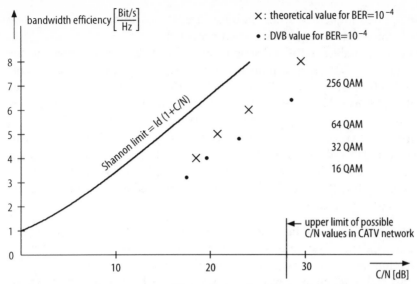

Fig. 10.13. Spectral efficiency of the types of modulation specified in the standard and the required signal-to-noise ratios at a bit error rate of $2 \cdot 10^{-4}$

10.6 DVB Utilsation in Master Antenna Television Networks

In the RACE Project "DIGISMATV" extensive work was carried out on the transmission of DVB signals in SMATV cable networks. On the basis of the obtained results the standard for the transmission of DVB signals in CATV networks [ETS 429] and that for the transmission of DVB signals in SMATV systems [ETS 473] were both adapted by the DVB Project. The results introduced here are based essentially on the findings by [DIGISMATV].

SMATV networks make available to their customers DVB signals which can be decoded with commercial IRDs. These must be transparent from the SMATV head-end right up to the user outlet, without the signal being converted to the baseband. Two different systems were defined to perform this task:

(1) System A includes the conversion of the DVB satellite signal (in accordance with [ETS 421]) into a DVB cable signal (in accordance with [ETS 429]). This conversion consists of a demodulation of the received QPSK signal and its Viterbi decoding as well as a subsequent remodulation by means of a quadrature amplitude modulator. The use of a de-interleaver, an RS-decoder and a unit for the energy dispersal as well as the corresponding processing steps at the encoder are not absolutely necessary. In [ETS 473] a differentiation is made between 'full implementation' and

'simplified implementation'. In both cases the customer requires a cable IRD for the decoding of the signal provided.

(2) System B enables a direct distribution of the DVB satellite signals in the cable network. Here, too, there are two variants, which depend on the intended frequency ranges. For the SMATV-IF system the signals in the satellite IF range (i.e. above 950 MHz) are fed into the distribution network. In this case the customer requires a satellite IRD in order to decode the signals delivered. In the SMATV-S system the satellite IF signal is downconverted to a range within the frequency band between 230 MHz and 470 MHz for distribution over cable. Before it can be decoded by a conventional satellite IRD decoder, it must be reconverted to the regular satellite intermediate frequency.

An essential factor in the choice of a SMATV system is the cost-performance balance. System B is the more economical solution. However, the application of the QPSK technique and the high transmission bandwidth that this technique entails exact a certain toll. The high frequency range of the satellite IF used in the SMATV-IF system for transmission in cable networks has the disadvantage of considerable cable attenuation, a fact which limits the system to short cable lengths. This disadvantage can be partially avoided by the SMATV-S system. However, in this case there is an additional requirement for frequency converters, which drives up the costs. It is only systems with a large number of subscribers that justify the expenditure entailed by the transmodulation, as recommended under (1). Numerous systems, however, do not attain the limit values for CATV networks discussed in section 10.1 because of the costs involved. Of particular importance are the differences in the domain of echo suppression. Investigations have shown that in SMATV systems there can occur echoes whose amplitudes differ considerably from those in CATV networks. Under adverse conditions they can have an amplitude which is only 10 dB below that of the main signal. Due to the cumulative effect of two time-shifted subsignals the frequency response assumes a ripple characteristic. The ripple r_m is expressed in logarithmic form as the ratio between maximum and minimum amplitudes:

$$r_m = 20 \log \left(\frac{1+\varrho}{1-\varrho} \right). \tag{10.22}$$

In the above case the ripple is approx. 6 dB.

The difference in delay time between the main signal and the echo results from the ratio of the echo path, which is twice the cable length, to the velocity of propagation of the wave in the cable. Depending on the type of cable used, the velocity of propagation is approx. 70% that of the speed of light, so that with a cable-connection length of 10 m the difference in delay time would be approx. 100 ns.

The distortions caused by the effect of the echo on the main signal can be cancelled by an equaliser at the receiver. Additional information is not required for the equalisation [BENVEN].

Simulations have shown that the transmission of a 64-QAM signal makes the use of an equaliser indispensable if the ripple exceeds 1 dB and if more than two ripples are to be found within the transmission channel. The deterioration of the signal as opposed to a transmission in an undistorted channel can be equalised up to 1 dB. Equalisers which can simultaneously process nine subsequent symbols will be quite adequate for this purpose. The equalisation of the channel impairment is mainly performed in the baseband so that the corresponding equalisers have a complex structure. The forms of implementation can nevertheless vary. In the literature the various algorithms are discussed in detail (e.g. [PROAKIS; LEE; KAMMEYER; GERDEN]).

10.7 Local Terrestrial Transmission (MMDS)

The DVB cable standard is intended not only for the transmission of DVB signals over cable but also for local terrestrial transmission in the microwave domain below 10 GHz. The systems operating in this microwave domain are called MMDS (microwave multichannel/multipoint distribution system, see section 9.6) and differ fundamentally from the traditional terrestrial broadcasting system by the designated frequency bands in which they operate and the related propagation characteristics of the electromagnetic waves. Distances of more than 60 km between transmitter and receiver, which are quite usual in traditional terrestrial broadcasting, are not possible in the microwave domain. The future technique, which can operate in the UHF and VHF bands in conventional transmission networks, is described in chapter 11. For transmission in the microwave region there must not only be short transmission paths, but also a direct line-of-sight from the transmitter to the receiver.

The DVB specification with the designation DVB-MC (DVB Microwave cable-based), has been ratified as European standard EN 300 749 [EN 749]. An important reason for adopting the cable standard was to ensure that identical signal processing would be used in as many applications as possible. A bandwidth-efficient and robust technique for modulation and channel coding was achieved with the cable standard and this should also be suitable for microwave distribution.

Symbols used in Chapter 10

A_K	MSB before differential coding and after differential decoding
B	bandwidth
b_i	bit of value $2i$
B_K	MSB-1 before differential coding and after differential decoding
C/I_{Ix}	carrier-to-interference ratio (as defined)
C/N_{Ix}	carrier-to-noise ratio (as defined)
E_b	energy per bit
f	frequency in general
f_{Ix}	frequency (as defined)
f_N	Nyquist frequency
$H(f)$	transfer function
$I, I(t)$	in-phase component of a signal
i	running variable, integer
I_K	MSB of the I-component after differential coding and before differential decoding
I_{K-1}	I_K delayed by one clock
M	number of constellation points of m bits which are transmitted per symbol, $M = 2^m$
m	bits per symbol or running variable
N	noise power
n	running variable, integer
N_o	noise-power density
PSD_{Ix}	power spectral density (as defined)
P_{Ix}	probability (as defined)
$Q, Q(t)$	quadrature component of the signal
Q_K	MSB of the Q-component after differential coding and before differential decoding
Q_{K-1}	Q_K delayed by one clock
R	code rate
r_m	in-band ripple
R_S	symbol rate, baud rate
S_{Ix}	signal power (as defined)
T_{Ix}	clock period (as defined)
T_S	symbol duration
α	roll-off factor
θ	phase rotation of echo channel
ϱ	amount of transfer function in the echo path
τ	delay time of echo

11 The Standard for Terrestrial Transmission and Its Decoding Technique

When comparing the stage of development of the standard for terrestrial transmission of DVB signals (DVB-T) with that of the standards for satellite and cable described in the previous chapters, one becomes easily aware that the terrestrial standard is less mature. There are some very good reasons for this. Firstly, the main priority for digital television was the transmission by satellite and cable, so it was only after completion of these two standards that a draft of the specification for terrestrial transmission was prepared, which was adopted in December 1995. Secondly, terrestrial transmission is far more complex than satellite or cable transmission, from the point of view of user requirements, with regard to the characteristics of the transmission path, and in view of the technical solutions that are called for. Furthermore, while the first two systems were being developed it was possible to draw upon practical experience acquired in the professional field, for instance with respect to modulation techniques for QAM and QPSK. The terrestrial transmission of digitised broadcasting signals, however, has only been tested in digital audio broadcasting up to now, and here, too, without the experience of having worked with large numbers of receivers. Apart from this, there had, until the end of 1998, only been test transmissions of DVB signals, namely in France, Germany, Great Britain and, conducted by the Grand Alliance, in the USA [IDSMITTEN, ENGELS]. In November 1998 the first – still growing – DVB-T network came into operation in the United Kingdom. During the same period a network was built up in the northern part of Germany, which by now includes some 20 operational transmitters.

The preparation of the specification which is presented here required the close co-operation of many dedicated and expert partners from various European countries. A prominent role was played, not only by individual industrial companies, broadcasting corporations and network operators, but also by the research consortia already mentioned in section 1.2.2.

11.1 Basics of Terrestrial Television Transmission

The distribution of television programmes by way of terrestrial transmitters is the "classical" technology of broadcasting. In contemporary analogue television in Europe (PAL, PALplus, SECAM) a typical content provider assumes the responsibility to provide a single television programme for the greatest possible section of the population. For example, in Germany this means that a typical viewer in a metropolitan area requires up to three different rooftop antennas in order to receive – at a good quality level – six or seven television programmes from up to three transmitting stations at different locations. "Quality" should be interpreted here as the overall effect, the sum of various influences including the many possible disturbances created by noise and interference. The concept of the provision of services to viewers and the extent of permissible disturbances are defined in the "Richtlinie für die Beurteilung der Fernsehversorgung bei ARD/ZDF und DBP", guidelines for the assessment of the provision of TV services from the first and second German public programmes and the German postal services (now Deutsche Telekom AG) [IRT].

A typical home receiver operates with a rooftop antenna which should have a gain of 10 dB (VHF) or 12 to 15 dB (UHF). The considerable directivity of the rooftop antenna, apart from the inherent increase in the signal power obtained by the gain, can partially reduce echo impairments caused by reflection from hills, buildings or suchlike. It is also possible to partially reduce the harmful effects of other transmissions in the same channel or in adjacent channels.

A good rooftop antenna guarantees television viewers in the service area of a transmitter a satisfactory video and audio quality, although some identifiable disturbances in the form of slight echoes, some noise and sporadic sputter can occur.

The reception conditions within the service area can, ideally, be described by the so-called Gaussian channel model, which is based on a direct signal path from transmitter to receiver, overlaid with additive white Gaussian noise (AWGN) which is mainly produced in the receiver itself.

In order to include the impairment caused by echoes it is necessary to enlarge the channel model. The transmission path known as "Ricean channel", takes into account the effect of multipath signals in addition to noise and to the dominating direct signal path between transmitter and receiver. The statistics of these multipath signals are approximated by a Ricean distribution [LÜKE 1]. As opposed to analogue television, in which echoes are more or less visible in the form of double contours on the screen, echoes in the case of DVB signal reception – when the delay time has exceeded a certain value – cause an increase in intersymbol interference (ISI), which ultimately results in an increase in the bit-error rate. This increase must then be corrected, for example, by increasing the transmission power. A simulation of the power requirements for the terrestrial DVB standard applied the following mathematical model to describe the Ricean channel, where 20 echoes were taken into consideration [ETS 744]. The output signal $y(t)$ of the channel model is described as a function of the input signal $x(t)$

$$y(t) = \frac{\varrho_0 \cdot x(t) + \sum_{i=1}^{N_e} \varrho_i \, e^{-j2\pi\Theta_i} \, x(t-\tau_i)}{\sqrt{\sum_{i=0}^{N_e} \varrho_i^2}} \qquad (11.1)$$

where:

ϱ_0 = attenuation in the direct signal path
N_e = number of echoes (here 20)
ϱ_i = attenuation in echo path i
Θ_i = phase rotation in echo path i
τ_i = relative delay time in echo path i

The Ricean factor $K = \varrho_0^2 / \sum_{i=1}^{N_e} \varrho_i^2$ denotes the ratio of the signal in the direct path to the sum of the signals in all echo paths. $K = 10$ dB was used for the simulation.

In Table 11.1 the values given for the simulation of the performance data for terrestrial DVB standards are for attenuation, relative delay time, and phase rotation. Figure 11.1 shows the attenuation of the 20 echo paths as a function of the echo delay time.

A comparison of the results of simulations, in terms of carrier-to-noise ratio (C/N) required for quasi error-free (QEF) reception of a DVB signal in the AWGN channel, with the respective values in the Ricean channel shows, as expected, that the Ricean channel has higher requirements. The additional requirement (see section 11.6) is actually in the range of 0,3 to 1.1 dB according to the choice of system parameters. This additional requirement is valid for a system with non-hierarchical modulation (see section 11.5). An increase in the transmission power by a maximum of 30% is therefore required to compensate the effect of echoes.

Reception with a rooftop antenna can be viewed as stationary reception and the directivity of the antenna can be used either for the selection of the direct signal or at least for choosing a dominant echo signal as the main reception signal. The situation changes in two main ways with the transition to a portable receiver. Firstly, the movement of the receiver during reception can cause changes to the reception conditions, particularly inside buildings. Secondly, rod antennas on portable receivers bring no noticeable antenna gain or directivity.

Table 11.1. Data for simulation of the terrestrial transmission channel. The values relate to the parameters in (11.1) and (11.2)

i	ϱ_i	τ_i [μs]	Θ_i [rad]
1	0.057662	1.003019	4.855121
2	0.176809	5.422091	3.419109
3	0.407163	0.518650	5.864470
4	0.303585	2.751772	2.215894
5	0.258782	0.602895	3.758058
6	0.061831	1.016585	5.430202
7	0.150340	0.143556	3.952093
8	0.051543	0.153832	1.093586
9	0.185074	3.324866	5.775198
10	0.400967	1.935570	0.154459
11	0.295723	0.429948	5.928383
12	0.350825	3.228872	3.053023
13	0.262909	0.848831	0.628578
14	0.225894	0.073883	2.128544
15	0.170996	0.203952	1.099463
16	0.149723	0.194207	3.462951
17	0.240140	0.924450	3.664773
18	0.116587	1.381320	2.833799
19	0.221155	0.640512	3.334290
20	0.259730	1.368671	0.393889

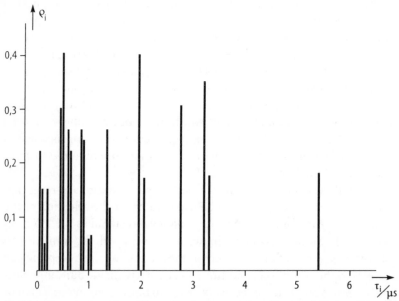

Fig. 11.1. Example of an echo profile used for the representation of the terrestrial transmission channel

The first version of the commercial and user requirements for terrestrial digital television in Europe defines the "stationary reception with a portable receiver" as one of its objectives. Mobile reception was thus deliberately not included as an objective. In comparison to the reception with a rooftop antenna, however, this limitation means it can no longer be taken for granted that a direct signal path will dominate. Consequently, for the simulation of the channel model which is representative of this case, the term which describes the direct signal path was eliminated in (11.1). This leads to (11.2).

$$y(t) = \frac{\sum\limits_{i=1}^{N_e} \varrho_i\, e^{-j2\pi\Theta_i}\, x(t-\tau_i)}{\sqrt{\sum\limits_{i=1}^{N_e} \varrho_i^2}} \tag{11.2}$$

The transmission channel modelled in this way, with echoes with more or less equal priority, is called "Rayleigh channel" [LÜKE 1] since the distribution of echoes corresponds to a Rayleigh distribution. Like the Ricean channel, the Rayleigh channel – as compared to the AWGN channel – requires a higher carrier-to-noise ratio. Section 11.6 presents simulation results which show that for an error-free reception – according to the choice of parameters in the standard – carrier-to-noise ratio is required which is up to 9 dB higher. This ratio, too, is valid in the case of non-hierarchical modulation (see section 11.5). An increase in the carrier-to-noise ratio by 9 dB would require an eight times higher transmission power. Naturally, an increase of that magnitude hardly seems realistic.

The portable receiver therefore is affected by the lack of gain in the receiving antenna (for example, 12 dB in the UHF band, which is the band most relevant to DVB), and by the signal deterioration in the Rayleigh channel. Additionally, further losses occur due to the fact that the reception antenna is often inside a building and often close to ground level, while an antenna height of 10 m from ground level is assumed in the calculation of the coverage according to [IRT]. For a combination of all these factors the ITU-R determined the following values, which should, however, be regarded as provisional [TELE-KOM]. In the case of a portable receiver outside a building the change of the antenna height from 10 m to 3 m results in a reduction in field strength by 6 dB. A further reduction in height from 3 m to 1 m would lead to a reduction by another 5 dB. Taking these values as a basis, the result is a requirement for an overall increase by approx. 30 dB (7–8 dB due to the Rayleigh channel, 12 dB loss in antenna gain, and 11 dB due to a decrease in height above ground level) in field strength for stationary reception with a portable receiver outside buildings. With reception inside buildings the field-strength loss is additionally increased owing to the attenuation caused by the walls of the buildings.

[TELEKOM] recommends that a calculated field-strength loss of 26 dB be assumed for the sum of the effects from wall attenuation and height decrease of the receiver position. In theory, the resulting overall effect requires an increase in field strength by up to about 45 dB (approx. 7–8 dB + 12 dB + 26 dB), if stationary reception using a portable receiver on the ground floor of a building is to be guaranteed.

Before implementation, the relevance of the above effects to practical applications must be verified in extensive field testing. In any case it seems safe to predict that coverage planning for DVB will not provide for field-strength levels of 45 dB to supply portable receivers. It may be assumed that stationary reception with portable receivers will become a by-product of the introduction of DVB. This means that, while portable receivers can be used close to transmitter stations, no reception will be possible with such equipment at an increasing distance from the transmitter and at a steadily increasing number of locations. This situation is quite different from that of the analogue terrestrial television transmissions, where video and audio quality deteriorate gradually. In the case of digital television the vision and sound fail abruptly at some critical point as the receiver is distanced from the transmitter or removed from an upper to a lower floor.

The hierarchical modulation described in section 11.5 will offer an option for the terrestrial transmission of DVB signals by which the sudden disruption of the reception, at least for individual programmes within a data container, can be mitigated.

Further it must be determined, from field trials of the system for the terrestrial transmission of DVB, which planning parameters for the protection ratios for rooftop reception and which antenna gain will actually be required. [PETKE]. An extreme case of such a determination could possibly be that an antenna gain of 0 dB is stipulated even for a rooftop antenna. Rooftop reception and stationary reception with a portable receiver would then only differ by the field-strength loss as a result of the reduction in the height of the reception position and by possible losses due to wall attenuation. The United Kingdom intends to introduce terrestrial DVB by retaining the existing rooftop antennas. Actually, the retention of existing antennas has come to be regarded as an important advantage for terrestrial transmission.

To summarise, even after the completion of the specification for terrestrial digital television (DVB-T), a multitude of unanswered questions remain, which will no doubt be addressed during the coming years.

11.2 User Requirements for a System for Terrestrial Transmission of DVB Signals

For the first time in the history of television engineering, terrestrial transmissions were not introduced first, the initial services being delivered by satellite or cable transmissions. Regarding the DVB Project, the systems for transmission by satellite and for subsequent distribution over cable have established many of the conditions for the standard of terrestrial television. The original user requirements of the system for terrestrial transmission of DVB signals have to be seen against this background. From the catalogue of user requirements only the most important criteria are listed below.

(1) The system for terrestrial transmission should have as much similarity as possible with the systems for cable and satellite. This will ensure that the home receiver technology will have as much similarity as possible with that for cable and satellite.

(2) DVB programmes should be transmitted in data containers and their capacities should be as large as possible. The channel bandwidth in Europe should be chosen so that a channel spacing of 8 MHz can be supported. Channels with 7-MHz spacing need not be supported, which implies that the introduction of DVB is not intended for bands I and III (VHF range).

(3) The system should have an optimum area coverage for stationary reception with a rooftop antenna. The support of stationary reception with portable receivers is desirable; mobile reception is not a development objective.

(4) It should be possible for DVB signals to be transmitted in terrestrial single-frequency networks. Single-frequency networks consist of transmitters which transmit exactly identical data streams in synchronism with each other, using the same transmission frequency. Neighbouring transmitters support each other in their coverage task. The topography of the network is essentially characterised by the maximum permissible distance to the next transmitter.

(5) The standard should allow the commencement of terrestrial transmission of DVB signals in 1997. The technology of the home receiver must be designed to enable the production of reasonably-priced equipment by 1997.

(6) Finally, "hierarchical modulation" should be included as an option.

For the technical layout of the system for the terrestrial transmission of DVB, requirements 1 and 4 are of paramount importance. In order to achieve the greatest possible similarity between the standards for terrestrial and satellite transmission the very same forward error correction was chosen which had been specified for satellite transmission. This comprises inner and outer error protection as well as interleaving (see sections 6.4 and 9.3). The requirement for the support of single-frequency networks leads automatically to the choice of an orthogonal frequency division multiplex (OFDM) as a modulation technique (see section 7.6). The combination of this modulation technique with the known methods of error protection has been termed "coded orthogonal frequency division multiplex" (COFDM).

Changes which were later incorporated in the commercial and user requirements were the result of non-European countries beginning to ask for solutions to 6-MHz and 7-MHz channel spacing and of the growing demand for mobile reception. As a consequence, DVB-T was eventually made available in versions appropriate for 6-MHz, 7-MHz and 8-MHz channel spacing and can even be received by fast-moving receivers.

For the design of a standard for the terrestrial transmission of DVB signals on the basis of COFDM it is the required maximum distance between mutually supporting transmitters in a single-frequency network which primarily decides what the minimum length of the required guard interval should be (see section 7.6). In simplified terms it can be said that the required duration of this guard interval can be computed if, for each receiver location at which it becomes impossible to decode the signal of one of two transmitters in the presence of signals from the other without the existence of the guard interval, the time difference between the two signals, computed by taking into account the distance from transmitter to receiver and the velocity of propagation, has been determined. The guard interval should then be longer than the greatest time difference resulting from this computation. In the United Kingdom, in particular, there are ongoing investigations into the relationship between the chosen length of the guard interval and the percentage of the population that can be provided with DVB services [LAFLIN]. According to these investigations 97% of the population can be supplied with a national single-frequency network if the length of the guard interval is 500 µs, 91% at 250 µs, 76% at 125 µs, and 63% at 62.5 µs. This analysis provides a possible approach to the determination of the length of the guard interval. When national single-requency networks need to be supported, the guard interval should be at least 200 µs. This represents a maximum difference of approx. 60 km in distances between neighbouring transmitters and the reception point.

In the process of generating the specification for the terrestrial transmission of DVB, the determination of the guard interval led to a particularly complex co-ordination problem between the participating nations. Probably none of the technical parameters with respect to DVB caused such prolific correspondence, even between the ministries responsible for broadcasting in the various countries, as did the length of the guard interval. It will be shown in section 11.3.2 that the choice of a guard-interval length of 200 µs represents a decision in favour of a rather complex variant of COFDM (8k COFDM). The adoption of this variant, particularly in the opinion of the representatives from the United Kingdom, meant that the complexity of the terrestrial receiver would be so great that the estimated costs for such equipment would be too high and the introduction schedule (1997) could never be met. Against the background of a White Paper published by the government in London in 1995 (see chapter 1) the relevance, in the United Kingdom, of a national single-frequency network seemed to have taken second place to the importance of the adherence to the implementation schedule. The representatives of other nations, such as Spain, strongly insisted on a guard interval with a length of 200 µs because of the belief that only with the introduction of national single-frequency networks terrestrial DVB would be possible in their countries. The Spanish partners of the DVB Project, as well as several others, attached less importance to the possible introduction of the system in 1997. As will be shown, it was possible to satisfy both positions by finally arriving at a specification which allowed for different complexities and at the same time provided for a choice of guard intervals with different durations.

11.3 Encoder Signal Processing

Numerous parts pertaining to the system for the terrestrial transmission of DVB signals are identical with those of the other systems previously described. Figure 11.2 depicts the block diagram of the encoder, in which the shaded components are exactly the same as those used in the satellite standard and therefore require no further explanation.

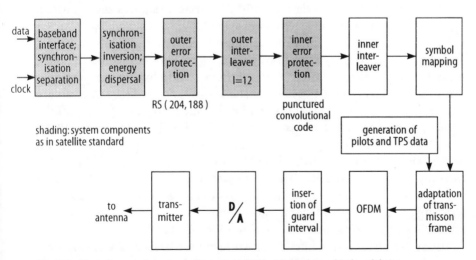

Fig. 11.2. Block diagram of an encoder for terrestrial DVB with non-hierarchical modulation

11.3.1 Inner Interleaver and Symbol Mapping

The first new processing element is an inner interleaver, which follows the inner error protection. Since OFDM is a multicarrier technique which can, among other things, be used to minimise the effects of frequency-selective impairments as a result of either multipath-reception interference or external sources of interference, the distribution of successive data over the large number of available carriers suggests itself. The distribution pattern should help disperse the effects of disturbances – even of a longer duration – affecting individual neighbouring carriers or a whole group of these, such that a correction will hopefully be achieved by the bitwise-functioning inner error protection.

The interleaving takes place in two steps. As a first step, 126 successive bits are combined into a block. Within this block the bits are then interleaved (bit interleaver). Subsequently a large number of blocks, generated from interleaved data in this way, is defined as a new block, within which whole groups of bits (symbols) are then interleaved (symbol interleaver).

It will be shown in sections 11.3.2 and 11.3.3 that in the two variants of the COFDM technique either 1512 (in 2k COFDM) or 6048 (in 8k COFDM) single carriers each are simultaneously modulated with useful data, and combined in an ODFM symbol. In addition to these 1512 or 6048 useful carriers another 193 or 769, respectively, are contained in the same OFDM symbol for synchronisation etc. The entire OFDM symbol therefore comprises either 1705 or 6817 single carriers. The process of interleaving refers to this structure of the OFDM symbol in various ways.

According to the choice of method for the modulation of the individual useful carriers with useful data, each of the carriers requires several useful bits. Alternative modulation techniques are QPSK, 16-QAM and 64-QAM. For QPSK, 2 bits are required for the modulation of each carrier (1 bit for the real and 1 bit for the imaginary axis in accordance with 4 possible constellations); for 16-QAM, 4 bits are required (2 bits for the real and 2 bits for the imaginary axis in accordance with 16 possible constellations) and for 64-QAM, 6 bits are required (3 bits for the real and 3 bits for the imaginary axis in accordance with 64 possible constellations). Consequently, for the creation of QPSK-modulated OFDM carriers two of the bitwise-functioning interleavers will also be connected in parallel. For 16-QAM there are four, and for 64-QAM there are six such interleavers. The interleaver structure for 16-QAM is shown as an example in figure 11.3. The choice of blocks of 126 bits for the bitwise interleaving makes it possible for a whole number of such blocks to be used in both COFDM variants ($1512 = 12 \cdot 126$, $6048 = 48 \cdot 126$).

The second level of the interleaver works on the basis of the symbols required for the modulation of the useful carriers. In order to make room for

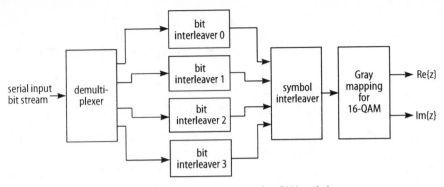

Fig. 11.3. Structure of an inner interleaver for the generation of 16-QAM symbols

the already-mentioned 193 or 769 carriers which are required for the synchronisation, the symbol interleaver generates no continuous stream of useful symbols at its output, but outputs an intermittent stream, which has intervals at points at which the additional carriers can be inserted.

In view of the great number of possible variations of the inner interleaving which depend not only on the modulation technique but also on the chosen COFDM variant, the possible types of interleavers cannot be dealt with at this point. For these the reader is referred to [ETS 744].

Each of the individual useful carriers of the OFDM signal is therefore separately modulated. A choice can be made between the modulation techniques QPSK, 16-QAM and 64-QAM. Furthermore, for the hierarchical-modulation technique (to be discussed in section 11.5) a so-called multiresolution QAM of type MR 16-QAM or type MR 64-QAM may also be selected.

The allocation of sequential bits to the symbols of the chosen modulation technique is carried out in accordance with a method named after Gray [LÜKE 1]. The reason for the selection of this technique is that it produces only one bit error in the case of the most probable transmission errors, i.e. when just one of the decision thresholds between two neighbouring constellation points is crossed. Consequently, the ordering of the symbols in the I/Q plane must be effected in such a way that the symbols running parallel to the I-axis and the symbols running parallel to the Q-axis may only differ from each other in one place. The result of the "Gray mapping" is shown in figure 11.4.

11.3.2 Choosing the OFDM Parameters

As already explained in section 7.6, in OFDM various parameters can be chosen separately to determine the performance of the whole system. One of the crucial parameters in the use of OFDM for DVB is the length of the guard interval. If single-frequency networks are to be realised in which the maximum

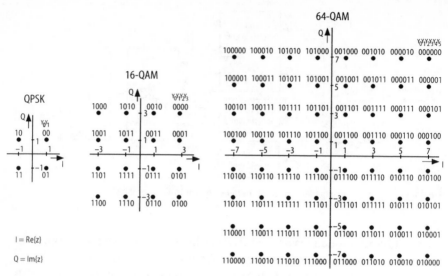

Fig. 11.4. Constellation diagram for the modulation of the individual carriers in the terrestrial DVB standard

permissible difference in distance between neighbouring transmitters and the corresponding receiver locations (further referred to, rather imprecisely, as transmitter spacing) is, for example, approx. 60 km, then the length of the guard interval T_g can no longer be chosen freely. This must then be at least 200 µs (200 µs · 300,000 km/s = 60 km). Now, the duration of the guard interval implies a reduction in the time available for the actual data transfer, which reduces the available channel capacity. Therefore the length of the guard interval is always kept relatively small as compared to the symbol duration T_S. On the other hand, long symbol durations automatically result in long durations of the useful part of the symbol T_U and therefore in reduced spacings between the individual carriers. Close carrier spacing, however, leads to complex circuitry in the receiver, which, after all, needs to be in a position to recover the information of the single carrier (see section 11.4). T_S might actually be fixed, for example, at 5 · 200 µs = 1 ms, and T_U would then be computed as 4 · 200 µs. From this follows that the spacing between the single carriers amounts to 1/800 µs = 1.25 kHz. On the basis of the chosen assumptions, 6000 neighbouring single carriers are transmitted in a UHF channel with a bandwidth of 8 MHz, of which, after deduction of protection zones at the band limits, approx. 7.5 MHz are effectively usable. On the assumption that each of these carriers is modulated with a 64-QAM, i.e. that 6 bits per carrier can be transmitted, the gross transmission capacity can be computed as (6000 · 6 bit per 1 ms =) 36 Mbit/s. Of this gross capacity, which also contains the required redundancy for the forward error correction, some portions must be reserved for synchronisation requirements etc.

In many applications the assumed guard interval of 200 μs is not appropriate. For example, if there is no necessity to create national single-frequency networks with complete coverage of the whole nation, but if, instead, only the coverage of a metropolitan area is required, it would be desirable to be able to work with shorter guard intervals in order to save channel capacity. Even when a transmission network is to be constructed which is based on conventional planning (an independent transmission frequency per transmitter site), the utilisation of a long guard interval is not desirable. For instance, when the transmitter spacing needs only to be in the range of 15 km, as in the case of the coverage of a metropolitan area by a single transmitter which is supplemented by low-power transmitters in the "canyons of houses" or on the outskirts of a coverage area, a guard interval of 50 μs is sufficient. On the analogy of the previous analyses this example would lead to a symbol duration T_S = 5 · 50 μs = 250 μs. T_U is then computed as 4 · 50 μs. In this way the spacing between the single carriers results in 1/(200 μs) = 5 kHz. Within one UHF channel approx. 1500 single carriers can be transmitted. When using 64-QAM for the modulation, the gross transmission capacity for each carrier again computes as (1500 · 6 bit per 250 μs =) 36 Mbit/s. Of this gross capacity, which of course includes the redundancy required for forward error correction, appreciable amounts must again be reserved for synchronisation and other purposes.

The study of the required length of the guard interval has shown that for terrestrial DVB large numbers of carriers are required, which in the case of national single-frequency networks would be in the range of 6000 and in the case of regional single-frequency networks, in the range of 1500. Since OFDM signals (see section 7.6) are generated by means of an IDFT, a solution would be to select, at the modulator, one-chip implementations of an IFDT which can carry out such a Fourier transform. These components, however, are so designed that the number of the samples used is equivalent to some power of two. The next power of 2 greater than 6000 is 2^{13} = 8192 (8k). The next power of 2 greater than 1500 is 2^{11} = 2048 (2k). The specification for terrestrial DVB had to be prepared for both types of SNFs due to the insurmountable differences of national interests explained in section 11.2. Therefore it not only includes a variant based on an 8k-IDFT but also one based on a 2k-IDFT. The working title chosen for this combination is "Common 2k/8k Specification".

On the basis of the two variants six possible values for T_g were agreed: 224 μs, 112 μs, 56 μs, 28 μs, 14 μs, and 7 μs. The four longer guard intervals belong to the 8k variants of the specification and the four shorter ones belong to the 2k variants. The reason for selecting precisely these values will be explained in the following. In any case, the order of magnitude shows that national SFNs can be realised with the longest guard intervals (transmitter spacing <67 km), that the middle range of values is conceived for regional networks (transmitter spacing 17 km or 33 km) and that the short guard intervals (transmitter spacing 2 km, 4 km or 8 km) are aimed at local coverage, where a master transmitter would for instance be supported by one or more gap-filling transmitters.

There now remains the sampling frequency to be determined for performing the IDFT. This was set at 64/7 MHz = 9.143 MHz. With this task completed, the time T_U is now clearly defined.

For the 8k variant, $T_U = 8192 \cdot (1/[64/7]$ MHz$) = 896$ µs. The spacing between the individual OFDM carriers can be computed as $(1/896$ µs$) = 1.116$ kHz. From this number it follows that not all 8192 carriers can really be transmitted. The actual bandwidth used has to be within the 8-MHz channel spacing and must remain limited to approx. 7.5 MHz. 6817 carriers were actually chosen within the limit of $k_{min} = 0$ and $k_{max} = 6816$. The OFDM signal therefore occupies a total bandwidth of $6817 \cdot (1/896$ µs$) = 7.609$ MHz.

For the 2k variant, $T_U = 2048 \cdot (1/[64/7]$ MHz$) = 224$ µs. The spacing between the individual OFDM carriers can be calculated as $(1/224$ µs$) = 4.4643$ kHz. It is logical that not all 2048 carriers can be transmitted, as the actual bandwidth used for this variant also has to be within the 8-MHz channel spacing and must remain limited to approx. 7.5 MHz. 1705 carriers were chosen within the limit of $k_{min} = 0$ and $k_{max} = 1704$. The OFDM signal therefore occupies a total bandwidth of $1705 \cdot (1/224$ µs$) = 7.612$ MHz.

The aim pursued in making the seemingly arbitrary choice of a sampling frequency of 64/7 MHz can now be explained. For if this value is reduced to (64/8 MHz) while retaining an 8k-IDFT, then this results in the parameters T_U = 1.024 ms, carrier spacing 977 Hz, and bandwidth 6.66 MHz. This means that after a simple change of the sampling frequency it is possible to use a 7-MHz channel if required. It goes without saying that another change to the appropriate value allows the use of DVB-T in 6-MHz channels ($64/7 \cdot 6/8$ MHz, T_U = 1.19 ms, carrier spacing = 837 Hz, bandwidth = 5.71 MHz).

The three values for the length of the guard interval can now be deduced from the time T_U (again only taking into account the specification for 8-MHz channels). For the variants 8k and 2k, these values are each at a ratio of $\Delta = (T_g/T_U) = 1/4$ or 1/8, 1/16 or 1/32.

The total symbol duration T_S then originates from the sum of the duration of the usable symbol and the duration of the guard interval, which in the case of the 8k variant is 1120 µs or 1008 µs or 952 µs or 924 µs. In the case of the 2k variant this results in 280 µs or 252 µs or 238 µs or 231 µs. On the above assumption that each single carrier is modulated with a 64-QAM, thus carrying 6 bits per carrier, the gross transmission capacity can be calculated, for example in the 8k variant and in the case $\Delta = 1/4$, as ($6817 \cdot 6$ bit per 1.120 ms =) 36.52 Mbit/s.

11.3.3 Arrangement of the Transmission Frame

The transmission of a complex signal such as the OFDM signal described above, comprising either 6817 or 1705 carriers modulated individually with QPSK or QAM with a spacing of about 1.1 kHz or 4.4 kHz, requires the addition of a considerable amount of synchronisation and signalling data. It is only due to this additional information that the receiver is enabled to utilise algorithms for certain control functions in order to realise an error-free demodulation of the signal. To keep the data rate sacrificed for the additional data within limits, it is sensible to combine a certain amount of the OFDM symbols and additional data to form a transmission frame. All the additional data relevant to that frame can then be transmitted only once. In the following, the term "symbol" denotes all 6817 or 1705 carriers which are transmitted simultaneously for the symbol duration T_S.

In the case of the specification for the terrestrial transmission of OFDM signals there are actually 68 symbols combined to form a transmission frame. Four such frames are defined as a superframe. Figure 11.5 depicts a section of the transmission-frame structure. The 6817 or 1705 subcarriers of the OFDM signal are shown next to each other in the horizontal direction. The abscissa therefore represents the frequency axis. The succession of the symbols is shown in the vertical direction.

Apart from the carriers which are modulated with the actual useful information, three further types of carriers which are required for the synchronisation or the transmission of additional information are used. Each of the carriers used for the synchronisation is individually modulated. The rules for the choice of a modulation word for each one of these carriers are known to the receiver.

The "continual pilots" are present in each symbol at the same carrier position (in the case of 2k, for instance, at the 45 positions 0, 48, 54, 87 ... 1323, 1377, 1491, 1683, 1704; in the case of 8k, for instance, at the 177 positions 0, 48, 54, 87 ... 6435, 6489, 6603, 6795, 6816). The distribution along the frequency axis is chosen so that within one channel no periodicities can occur. Therefore the continual pilots can be used for the coarse adjustment of the frequency of the local oscillator in the receiver. In order to make the synchronising information as robust as possible against transmission errors, the continual pilots have an amplitude which, in comparison to the carriers containing useful information, is increased by the factor 4/3. That is why they are also called boosted pilots.

The "scattered pilots" are scattered over the transmission frame according to a definite plan. This plan can be easily identified in figure 11.5. The aim of the scattering is, firstly, to make available a large number of pilots within each symbol for the fine adjustment of the receiver. Secondly, the scattered pilots

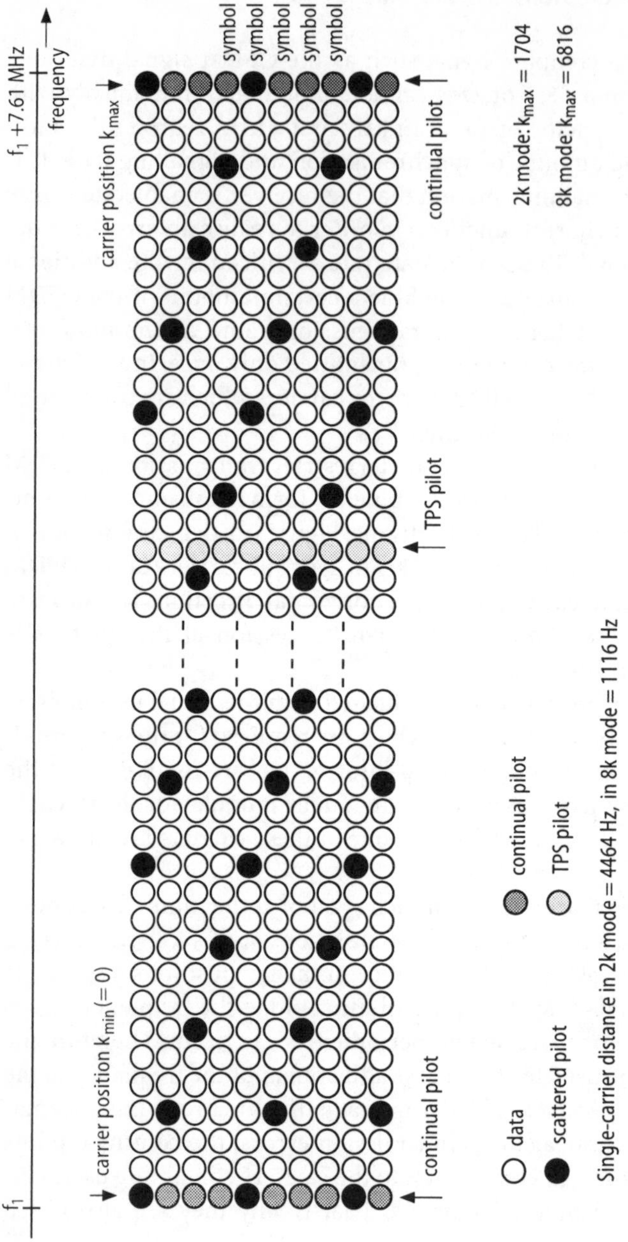

Fig. 11.5. Transmission frame for a COFDM signal

- ○ data
- ● scattered pilot
- ◉ continual pilot
- ◯ TPS pilot

Single-carrier distance in 2k mode = 4464 Hz, in 8k mode = 1116 Hz

2k mode: $k_{max} = 1704$
8k mode: $k_{max} = 6816$

$f_1 + 7.61$ MHz

frequency

carrier position k_{max}

continual pilot

TPS pilot

symbol 67
symbol 0
symbol 1
symbol 2
symbol 3

carrier position $k_{min} (= 0)$

continual pilot

f_1

can contribute to the temporal synchronisation. Thirdly, the receiver will be able to perform a nearly continuous analysis of the frequency- and the temporal plane, which can be used for the evaluation of the current channel condition ("channel estimation"). The scattered pilots are also transmitted with increased amplitude.

The "transmission-parameter signalling pilots" (TPS pilots) permit a transmission of additional information with which the temporal synchronisation, i.e. the recognition of the beginning of the transmission frame, can be ensured. Moreover, they provide the information about the transmission parameters used. The details of this will be discussed in the following. The TPS pilots, just as the continual pilots, can be found in precisely fixed carrier positions (in the case of 2k, for instance at the 17 positions 34, 50, 209, 346, 413 ... 1262, 1286, 1469, 1594, 1687; in the case of 8k, at the 68 positions 34, 50, 209, 346, 413 ... 6374, 6398, 6581, 6706, 6799). TPS pilots are transmitted with an amplitude which represents the average amplitude of all carriers transmitting useful information.

The numbers of the various types of pilots named per symbol cannot just be added to determine the channel capacities lost for the transmission of useful information, as the possibility that a continual pilot may be at a position intended for a scattered pilot cannot be ignored (see figure 11.5). On closer examination it may be seen that each of the continual pilots is co-located in every fourth symbol with a scattered pilot. This results, on average, in 1512 carriers with useful information per symbol in the 2k case, and in 6048 carriers with useful information in the 8k case, which means that approx. 13% of all carriers have to be utilised for synchronisation, etc.

Attention has already been drawn to the fact that the continual pilots and the scattered pilots are individually modulated so that from an analysis of this modulation the receiver can, among other things, derive information for the tuning. The modulation is carried out in accordance with the structure of a maximum-length sequence which is known to the transmitter and to the receiver.

Maximum-length sequences (m-sequences) are often referred to as pseudo-random sequences. They have some prominent features. One of these is the ideal, pulse-shaped, periodically recurring auto-correlation function which is of particular importance in this connection [LÜKE 2].

A maximum-length sequence can be generated by a shift register in which the positions indicated by the descriptive polynomial are linked by means of EX-OR circuits (s.a. section 9.3). The polynomial used here has the form

$$g(X) = X^{11} + X^2 + X^0 .$$

X^{11} represents the input of the first cell, X^0 represents the output of the shift register. All sequences commence with the initialisation condition $X^s = 1$, in such a way that the first bit at the output of the shift register coincides with the first carrier of a symbol. Further shifting within the register takes place such that there is a shift step to each carrier, independently of whether it is a pilot or not. When the temporally coincidental carrier happens to be a continual pilot or a scattered pilot, a modulation symbol is allocated to the pilot by the bit at the output of the shift register. When this bit is 0, then the pilot is modulated with the positive real part $+4/3$; when the bit is 1, then the pilot is modulated with the negative real part $-4/3$. The imaginary part of the modulation symbol is always zero. If the adjacent carriers in figure 11.5 which form part of a symbol are examined from left to right, i.e. along the frequency axis, a subsampled maximum sequence will be detected at the 176 or 701 positions per symbol at which continual or scattered pilots are to be found. This maximum sequence can now be used for fine-tuning in the receiver.

In every transmission frame the data denoted as TPS are transmitted within each symbol. Basically, they constitute a special data channel which can carry important information for the duration of every transmission frame. The receiver requires this information for the demodulation of the subsequent transmission frame. The purpose of the TPS data is therefore similar to that of the so-called fast-information channel of the DAB system [ETS 401]. Within each transmission frame there are 68 successive symbols. Each of these symbols contains one bit from a 68-bit TPS word. For the 2k variant 17 carriers per symbol are made available for TPS, and for the 8k variant 68 carriers. In each symbol these TPS carriers all transmit exactly the same individual bits in parallel.

The information made available by the TPS refers to the modulation technique (QPSK, 16-QAM, 64-QAM), the constellation pattern in the case of hierarchical modulation with the possible values $\alpha = 1$ or $\alpha = 2$ or $\alpha = 4$ (see section 11.5), the code rate of the inner error protection $(1/2, 2/3, 3/4, 5/6, 7/8)$, the duration of the guard interval $(\Delta = 1/32, \Delta = 1/16, \Delta = 1/8, \Delta = 1/4)$, the OFDM variants (2k and 8k), and the position of the transmission frame within the superframe (1, 2, 3, 4).

In view of the importance of an error-free reception of the TPS data, these are transmitted in parallel many times over and are also separately error-protected. A shortened BCH code (67, 53) is used. The individual bits are modulated on the TPS carriers with differential 2-PSK (see section 7.2).

A maximum of 68 bits can be used for TPS. 31 bits are required for word recognition and error protection. 23 bits are used at the moment. The remaining 14 bits are reserved for future use.

11.4 Decoding Technique

An essential condition for designing a receiver for OFDM signals with 6817 carriers is the availability of a DFT circuit component capable of handling 8192 samples and with sufficient processing speed. The first prototypes of such an integrated circuit were built in the summer of 1994 [BIDET]. They are based on a 0.5 μm CMOS technology and at that time represented a truly admirable feat of engineering. However, due to the high costs their use in mass-produced home receivers was out of the question. Nevertheless there was general agreement that only after the introduction of the 0.35 μm CMOS a cost-level would be reached which would allow a large-scale manufacture of receivers. It was expected that this technology would support the integration of all the digital functions which are necessary for the demodulation and the decoding of the signal into a chip measuring less than one square centimetre. However, the question was not as to when this technological improvement would be available in the laboratory, but rather when it could be ready for the mass production of integrated circuits. In the meantime, the one-chip DVB-T decoder has become a reality.

The stability of the oscillators required in the receiver is of great importance for the overall performance. On the one hand, the single carriers which are approx. 1.1 kHz or 4.4 kHz apart can only be separated free of crosstalk when it can be guaranteed that the sampling is carried out exactly at the zero crossings of all $\sin(x)/x$ functions not belonging to the desired carrier (see section 7.6). On the other hand, it must remain possible for the phase information which is integrated in the QPSK- or QAM-modulated individual carriers to be recovered. With the aid of the various auxiliary data described in section 11.3 it is possible to accurately set the oscillator frequency and also the static phase of the local oscillator in the receiver. However, these static quantities are superimposed by phase noise. This results, first of all, in a phase error which affects all the individual carriers within a symbol in the same way, but which shows little correlation with the phase errors within the subsequent symbols. Secondly, the phase noise can lead to phase errors which affect different carriers within the same symbol in different ways. This is why both the level and the spectral distribution of the phase noise in the local oscillator are of interest. Fortunately, first experiments with concepts for oscillators which are very similar to those used in current home receivers showed that it is possible to master the problem of phase noise with the help of conventional technology [MUSCHALL]. Comparison measurements of the bit-error rate with and without the influence of a local oscillator showed only a minimal difference. By increasing the carrier-to-noise ratio in the transmission channel by approx. 0.5 dB the deterioration of the bit-error rate as a consequence of the phase noise could be fully compensated.

The measuring of the channel frequency response for attenuation and phase can be carried out on the basis of the continual pilots and on the basis of the scattered pilots. A compensation of the frequency response based on the measuring results can be used to correct level and phase of each single carrier. Moreover, a reliability statement can be derived from the measurements, which must be made available to a soft-decision Viterbi decoder. When the DVB-T system was developed it was not exactly known which residual errors would remain after the compensation, for example as a result of a less-than-perfect measurement of the channel-frequency response, nor could the magnitude of the expenditure for measurement and correction in the receiver be determined. Inasmuch as the effects of the residual errors could also be compensated by an increase in the carrier-to-noise ratio in the transmission channel, no statement could be made about a nominal loss of the carrier-to-noise ratio.

To sum up the then unresolved questions regarding the decoding technique in the receiver, there was no assurance that the requirements regarding the timing of the introduction of home receivers and the cost of first-generation receivers (see requirement no. 5, section 11.2) could have been met if solely the 8k variant had been specified, as originally planned in early 1995. According to nearly all member organisations of the DVB Project, the inclusion of the 2k variant represented a technical solution that ensured the introduction of terrestrial transmissions and the sale of cost-effective receivers as early as 1997.

The question arose as to the compatibility between the two variants. Would a 2k-based receiver, sold in the UK in 1997, also demodulate those transmissions which may start in Spain in the year 2000 on the basis of the 8k variant? The answer to this question is clear. Due to the use of the less expensive 2k IDFT circuit in the receivers these will not be suitable for reception of 8k signals. On the other hand, receivers designed for the 8k option will also be able to demodulate the 2k signals, provided they are manufactured in accordance with the DVB specifications. This means there is a possibility that, say, from the year 2000 only one type of receiver will be marketed in the whole of Europe which will be capable of handling both 2k and 8k.

11.5 Hierarchical Modulation

It is a fact that all DVB transmission systems exhibit an abrupt failure characteristic. A variation in the carrier-to-noise ratio of only a fraction of a dB can cause a complete breakdown of the transmission path. This effect is, of course, of less importance in cable systems. A signal of sufficient quality can always be guaranteed in cable networks when the failure criteria have been taken into account during the planning of the network. A fluctuation of the reception conditions is not to be expected. In satellite transmission, fluctuations in the carrier-to-noise ratio can be caused by changes in the meteorological conditions. The use of sufficiently-sized reception dishes (diameter ≥ 60 cm) directed exactly at the satellite, should ensure that there are very few moments of breakdown in the course of a year (see chapter 9). Weather conditions can affect a wide area causing widespread failure. A storm front traversing a region, for instance, can cause a breakdown in many places at the same time. During good weather conditions and with a suitable receiving installation every viewer can receive the programmes on offer free of transmission errors. This possibility only exists, of course, if the antenna is installed in line-of-sight with the satellite.

Similar conditions as those described for satellite reception will prevail for the reception of terrestrially transmitted DVB signals with a rooftop antenna. Those viewers who are able to receive sufficiently high-powered signals, either directly or as echoes, will receive their choice of programmes free of transmission errors. Fluctuations in the carrier-to-noise ratio can also occur in terrestrial transmissions, for instance due to seasonal changes in the echo impairments or due to aeroplanes flying past. However, these are far less drastic than the meteorologically induced disturbances in satellite systems. The real difference between terrestrial and satellite reception is characterised, on the one hand, by the objective for terrestrial transmitters to also facilitate, within certain limits, the coverage of stationary portable receivers with a rod antenna and, on the other hand, by the expectations of the viewers. The viewer of analogue TV programmes is used to, and will therefore expect, a flawless reception of at least the public television programmes, provided he observes certain rules concerning the type and height of the rooftop antenna required at his place of residence. Reception on portable receivers equipped with set-top aerials is regarded as a bonus. These viewer expectations can only be met at considerable expenditure on the part of the broadcasters, even allowing for the "graceful degradation" of analogue signals. DVB signals with their abrupt failure behaviour complicate the situation considerably. For example, it is conceivable that while large areas of a city might be well served with terrestrial DVB signals, one district might be situated on lower-lying ground where no reception whatsoever is possible. To supply this district, a general increase in the transmission power would of course be possible, and/or instead of each single carrier being modulated with 64 QAM, a QPSK modulation could be applied and/or a lower code rate for the inner error protection could be utilised. The reception could also be improved by using fill-in transmitters. All these procedures have one distinct disadvantage; they would result in a less efficient use of the available channel capacity while at the same time only few viewers might profit from the above measures.

The hierarchical transmission technique is based on the division of the television channels used for the terrestrial transmission of DVB signals into two parts. The first part enables relatively low data rates (a few Mbit/s) to be transmitted in such a way that they can still be received in the case of relatively poor carrier-to-noise ratios. Parallel to this, the second part enables the transmission of considerably higher data rates, however with higher carrier-to-noise ratio requirements. The concept of the "data container" can be maintained if "hierarchical transmission technique" is to be understood as dividing the data container into a robust and a fragile portion. Both "subcontainers" are nevertheless transmitted in the same channel. Within the robust portion of the data container, for example, some basic (public) programmes could be transmitted (although with a somewhat more modest audio and video quality) and the less robust portion could accommodate new, additional programmes. Alternatively, the basic programmes could be transmitted once again, but with a considerably improved audio and video quality. Due to the commercial and user requirements described in section 11.2 the standard for terrestrial transmission of DVB signals does not include any mechanisms for the so-called hierarchical source coding. Had that been implemented, the robust portion of the data container could be used to transmit a programme in modest quality, while the less robust portion of the container could transmit additional data for the same programme, which could then exploit all the data of the entire container in order to enhance that programme to, for example, HDTV quality.

A concept for the hierarchical transmission is easily defined. Whilst retaining the normal transmission frame it should be possible to transmit two data streams at the same time; one modulated with QPSK and with a lower code rate for the inner error protection, the other modulated with either 16-QAM or 64-QAM but with a higher code rate for the inner error protection. The necessary interleaving of the two data streams could either be designed as a temporal multiplex or a frequency multiplex or even as a modulation multiplex. The hierarchical modulation in the intended form utilises the modulation-multiplex method. This means that each single carrier in the transmission frame transmits two data words simultaneously, of which one comprises a high-priority data stream which needs to be specially protected, and the other a lower-priority data stream with less need for protection.

Figure 11.6 shows a part of the block diagram of an encoder for terrestrial DVB with hierarchical modulation. A splitter divides the input data stream. Both partial streams subsequently pass through identical processing steps; however, the inner error protection is chosen differently in each case. Subsequent to the inner interleaver the two partial streams are reunited in the inner interleaver and then further processed as shown in figure 11.2.

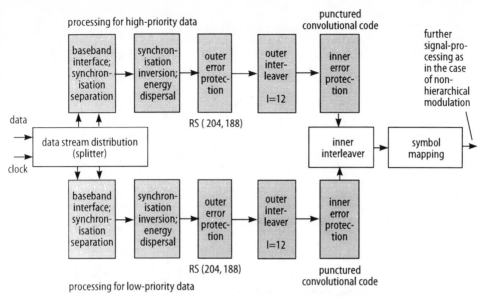

Fig. 11.6. Block diagram of an encoder for terrestrial DVB with hierarchical modulation

The reuniting of the partial streams is outlined in figure 11.7. The case of the generation of a 64-QAM symbol with six bits per symbol is shown as an example. Two of the six bits come from the high-priority data stream and the other four bits are from the low-priority data stream. As an alternative, it would also be possible to generate a 16-QAM symbol with four bits per symbol, with two of the four bits coming from the high-priority data stream and two from the low-priority data stream. In both cases the essence of the mapping lies in the fact that the two bits which come from the high-priority data stream (y_0, y_1) define the quadrant in which the symbols are to be found (figure 11.8). Hence it could be said that hierarchical modulation enables the em-

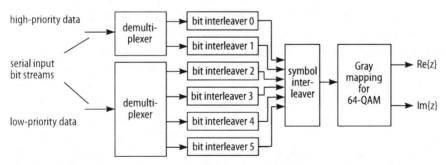

Fig. 11.7. Structure of an inner interleaver for the generation of 64-QAM symbols with embedded QPSK symbols

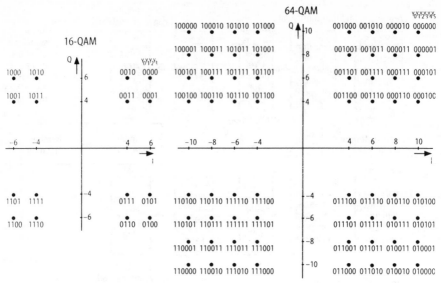

Fig. 11.8. Constellation diagram for the hierarchical modulation ($\alpha = 4$)

bedding of a QPSK symbol, as defined by the high-priority data, in either a 16-QAM or a 64-QAM symbol. The term "multiresolution QAM" adequately describes the situation.

Figure 11.8 could be found puzzling in that the constellation points within each quadrant have a different spacing from the constellation points in neighbouring quadrants. The spacing measure is defined by the characteristic quantity, the value of which can also be transmitted within the TPS (see section 11.3). The values 1, 2, and 4 are allowed for. Figure 11.4 shows the condition for $\alpha = 1$. In figure 11.8 the constellation for $\alpha = 4$ is shown. It can be seen by a comparison of the two diagrams that α represents the ratio between the spacing of two adjacent constellation points of two neighbouring quadrants and the spacing of the constellation points within one quadrant.

For a given type of QAM, an increase in α leads, on the one hand, to an increase in the average transmission power which follows from the extension of the length of the vector between the origin of the ordinates and the mean constellation point. For a given transmission power, on the other hand, an increase in α leads to an increase in the required carrier-to-noise ratio for the demodulation of the QAM symbols, namely to an increase by some 4 dB at each doubling of α. Moreover, from figure 11.8 it is obvious that the permissible noise for an error-free demodulation of the QPSK component increases with α. Hence it can be said that an increase in α renders the robust portion of the data container even more robust, while at the same time the less robust portion becomes more error-prone.

The implementation of a receiver which is able to process hierarchical modulation can be carried out in two ways. A simple receiver differs from a receiver for non-hierarchical transmission only in that the demodulator/demapper must be able to split off either a high-priority data stream or a low-priority data stream, in accordance with the information about the type of modulation (that is, the value of α and the inner code rate continuously transmitted in the TPS), and must then forward the data stream to the (single-channel) channel decoder. The disadvantage of using this simple receiver technology is that in the case of a deterioration of the reception conditions the receiver needs to switch from the low-priority data stream over to the high-priority data stream, and this causes a short-term disruption in the continuity of the audio and video. This disruption is caused by the fact that the receiver needs to adapt to the new reception parameters. Alternatively, a more expensive receiver is conceivable, which is designed with two channels after the demodulator/demapper and which enables an immediate switching of at least the channel decoder. The effects of the switching of the source decoder, however, can only be avoided if two such components are operated in parallel.

11.6 Features of the Standard

The performance of the system for the terrestrial transmission of DVB signals can be analysed according to two criteria: (1) the available useful data rate and (2) the carrier-to-noise ratio in the transmission channel which is required for QEF reception.

However, the analysis of the performance data can lead to very intricate results. This is due to the vast number of parameters that can be chosen. It should be remembered that all of the following parameters can be selected independently of one another: the 2k and the 8k variant, the code rate of the inner error protection, the type of modulation of each single carrier (with the further option of a choice of either hierarchical or non-hierarchical modulation), the length of the guard interval and, finally, in the case of hierarchical modulation, the value of α. The required carrier-to-noise ratio in the transmission channel has to be investigated for the three channel models (Gaussian, Ricean and Rayleigh). It is evident from the above that in this paragraph a limitation to just a few important combinations is unavoidable.

11.6.1 Determination of Useful Data Rates

The basis for the determination of the useful data rates is the determination of the number of packets of the MPEG-2 transport stream that can be accommodated in a transmission frame. In this, the fact must be taken into account

Table 11.2. Number of transport packets in one superframe

Code rate	QPSK "2k"	"8k"	16-QAM "2k"	"8k"	64-QAM "2k"	"8k"
1/2	252	1008	504	2016	756	3024
2/3	336	1344	672	2688	1008	4032
3/4	378	1512	756	3024	1134	4536
5/6	420	1680	840	3360	1260	5040
7/8	441	1764	882	3528	1323	5292

that the outer error protection turns 188-byte packets into 204-byte packets by the addition of 16 bytes redundancy. The inner error protection then adds further redundancy according to the code rate. One should think that it is impossible to accommodate a whole number of transport packets in one transmission frame for all possible code rates and all possible types of modulation. Stuffing data seem to be unavoidable. But in point of fact it was possible to define the capacity remaining for useful data within a transmission frame such that for all conceivable combinations complete transport packets can be accommodated in a remaining superframe consisting of four transmission frames. The numerical background has already been considered in section 11.3. It was shown that for the 2k variant 1512 carriers per symbol remain for the useful data, and for the 8k variant, 6048. In a superframe this would be 272 · 1512 = 411,264 carriers and 272 · 6048 = 1,645,056 carriers respectively. Since, depending on the chosen kind of modulation of the single carriers, 2 bits per carrier (QPSK), 4 bits per carrier (16-QAM), or 6 bits per carrier (64-QAM) are transmitted, at least 272 · 1512 · 2 = 822,528 bits will fit into a superframe. If this "elementary quantity" is divided by (8 · 204 =) 1632 (the number of bits of a transport packet with error protection), the result is 504, which is a multiple of 8, 6, 4, 3, 2, i.e. of the divisors of the puncturing rates of the inner error protection.

Table 11.2 documents the number of 204-byte packets in both variants as a function of the code rate of the inner error protection and the chosen modulation.

It can easily be seen that, for 16-QAM, the number of transport packets per transmission frame is always double that for QPSK, and that for 64-QAM it is three times that for QPSK. In the case of QPSK a gross total of 822,528 bits, including the bits for the outer error protection – as already computed using the number of carriers available for the transport of useful data in a superframe – can be conveyed per superframe with the 2k variant. This value is computed, for example in the case of code rate 1/2, from the relation (204 bytes per transport packet) · (8 bits/byte) · (1/code rate) · 252 packets = 822,528 bits. With this method it is easy to compute the net data rates if the duration of the superframe is known. The net data rates should be interpreted as referring to

Table 11.3. Useful data rates for non-hierarchical modulation [Mbit/s]

Modulation	Code rate	Relative length of guard interval (Δ)			
		1/4	1/8	1/16	1/32
QPSK	1/2	4.98	5.53	5.85	6.03
	2/3	6.64	7.37	7.81	8.04
	3/4	7.46	8.29	8.78	9.05
	5/6	8.29	9.22	9.76	10.05
	7/8	8.71	9.68	10.25	10.56
16-QAM	1/2	9.95	11.06	11.71	12.06
	2/3	13.27	14.75	15.61	16.09
	3/4	14.93	16.59	17.56	18.10
	5/6	16.59	18.43	19.52	20.11
	7/8	17.42	19.35	20.49	21.11
64-QAM	1/2	14.93	16.59	17.56	18.10
	2/3	19.91	22.12	23.42	24.13
	3/4	22.39	24.88	26.35	27.14
	5/6	24.88	27.65	29.27	30.16
	7/8	26.13	29.03	30.74	31.67

the 188 net bytes in each transport packet. Of course, such a calculation includes the respective symbol duration T_S and therefore the length of the chosen guard interval. Let the data rate be calculated by way of an example (2k, QPSK, code rate 1/2, $\Delta = 1/4$ corresponding to $T_S = 280$ µs). The calculation leads to $822{,}528$ bit $\cdot$ $(1/2)$ $\cdot$ $(188/204)$ $\cdot$ $[1/(4 \cdot 68 \cdot 280$ µs$)] = 4.98$ Mbit/s. If this is changed to the 8k variant (8k, QPSK, code rate 1/2, $\Delta = 1/4$ corresponding to $T_S = 1120$ µs), the calculation of the data rate leads to (204 byte per transport packet) $\cdot$ (8 bit/byte) $\cdot$ (1/code rate) $\cdot$ 1008 packets $\cdot$ $(1/2)$ $\cdot$ $(188/204)$ $\cdot$ $[1/(4 \cdot 68 \cdot 1120$ µs$)] = 4.98$ Mbit/s. In other words, the data rate is identical to that in the 2k variant. This result, which is somewhat surprising at first glance, is explained by the fact that for the 8k variant four times the number of transport packets are transmitted as for the 2k variant (s.a. table 11.2), whereas at the same time the symbol duration is also exactly four times as long with the same relative guard interval (Δ). All possible data rates are given in Table 11.3.

Using table 11.3 it is also possible to determine all the various useful data rates for the different variants of the hierarchical modulation. For example, to determine which useful data rate results from hierarchical modulation in the case of "QPSK embedded in 16-QAM", the data rate for the high-priority data stream is to be found in the corresponding line of the block of figures relative to QPSK, in accordance with the chosen error protection and the Δ used. The data rate for the low-priority data stream can also be found in the corresponding line of the block of figures relative to QPSK, in accordance with the chosen error protection and the Δ used. In the case of hierarchical modulation of the type "QPSK embedded in 64-QAM" the data rate for the high-

Table 11.4. Bandwidth efficiency for non-hierarchical modulation in the case Δ = ¼

Code rate	QPSK [bit/s per Hz]	16-QAM [bit/s per Hz]	64-QAM [bit/s per Hz]
1/2	0.62	1.24	1.87
2/3	0.83	1.66	2.49
3/4	0.93	1.87	2.80
5/6	1.04	2.07	3.11
7/8	1.09	2.18	3.27

priority data stream is to be found in the QPSK block of figures for the relevant combination of code rate and Δ; and the data rate for the low-priority data stream, in the 16-QAM block of figures for the relevant combination of code rate and Δ. If in this last example, for instance, the high-priority data stream is protected by an inner error protection with the code rate 1/2, and the low-priority data stream is protected by an inner error protection with the code rate 5/6, then the data rate for the high-priority data stream is 4.98 Mbit/s. To this value, for the accompanying low-priority data stream, a data rate of 16.59 Mbit/s is added. The result is a total data rate of 21.57 Mbit/s.

It might be instructive to calculate the limits of the values possible for the bandwidth efficiency introduced in the standard. For this the respective data rate is divided by the channel bandwidth (8 MHz). This results in values in the form bit/s per Hz. In table 11.4 these values are represented for Δ = 1/4.

Furthermore it can be seen from table 11.3 that a decrease in the relative guard interval leads to a corresponding increase in the data rate. Considering the bandwidth efficiency, this means that, taking the example of a 64-QAM modulation and a code rate of 7/8, the bandwidth efficiency increases from 3.27 bit/s per Hz in the case of Δ = 1/4 to 3.96 bit/s per Hz in the case of Δ = 1/32. Regarding these values for the bandwidth efficiency, it should be noted that the loss of efficiency caused by the unavoidable bandwidth limitation and by the insertion of reference signals, TPS data, error protection, etc. has been fully taken into account.

11.6.2 Required Carrier-to-noise Ratio in the Transmission Channel

The values given in the following table are based on simulations of the system behaviour. They were computed on the assumption that a perfect correction of the channel frequency response had taken place in terms of attenuation and phase rotation. Phase noise, as a source of errors within the receiver, was considered to be non-existent in these simulations. As a result of these idealised assumptions the behaviour of the 2k variant no longer differs from that of the 8k variant. As with the investigations into the characteristics of the standards for transmission over satellite and cable, a carrier-to-noise ratio

Table 11.5. Minimum C/N ratio in the transmission channel in the case of non-hierarchical modulation required for QEF reception

Type of modulation	Code rate	Gaussian channel [dB]	Ricean channel [dB]	Rayleigh channel [dB]
QPSK	1/2	3.1	3.6	5.4
	2/3	4.9	5.7	8.4
	3/4	5.9	6.8	10.7
	5/6	6.9	8.0	13.1
	7/8	7.7	8.7	16.3
16-QAM	1/2	8.8	9.6	11.2
	2/3	11.1	11.6	14.2
	3/4	12.5	13.0	16.7
	5/6	13.5	14.4	19.3
	7/8	13.9	15.0	22.8
64-QAM	1/2	14.4	14.7	16.0
	2/3	16.5	17.1	19.3
	3/4	18.0	18.6	21.7
	5/6	19.3	20.0	25.3
	7/8	20.1	21.0	27.9

(C/N) was determined at which the bit-error rate – after decoding the inner error protection – is equal to or less than $2 \cdot 10^{-4}$. As explained in section 6.3.5, this condition leads to practically error-free signals (QEF) after the Reed-Solomon decoder. A compilation of the required C/N ratio (in dB) for all possible combinations of modulation techniques and code rates of the inner error protection is given in table 11.5. The data refer to non-hierarchical modulation.

A close analysis of the numerical values in table 11.5 leads to the following conclusions. A change from QPSK to 16-QAM with a constant code rate results in an increase in the required C/N by approx. 6 dB. The same applies to the change from 16-QAM to 64-QAM. The transition from the Gaussian channel to the Ricean channel, with the same type of modulation and constant code rate, results in a necessary increase in the C/N by a maximum of 1.1 dB. As explained in section 11.1, the simulation of the Ricean channel given by (11.1) represents an approximation of the actual conditions when receiving DVB signals via a rooftop antenna with a high directivity. The transition from the Gaussian channel to the Rayleigh channel, with the same modulation type and the same code rate, results in a necessary increase in the required C/N by a maximum of 8.9 dB. The Rayleigh channel given by (11.2) is used to model the actual conditions when receiving DVB signals via stationary receivers which have a rod antenna.

In the case of hierarchical modulation it is necessary to supply details of the required C/N not only for the QPSK portion but also for the QAM portion.

Table 11.6. Minimum C/N in the transmission channel for hierarchical modulation in the form QPSK embedded in 64-QAM ($\alpha = 2$) required for QEF reception

Type of modulation	Code rate	Gaussian channel [dB]	Ricean channel [dB]	Rayleigh channel [dB]
QPSK	1/2	6.5	7.1	8.7
	2/3	9.0	9.9	11.7
	3/4	10.8	11.5	14.5
in				
64-QAM	1/2	16.3	16.7	18.2
	2/3	18.9	19.5	21.7
	3/4	21.0	21.6	24.5
	5/6	21.9	22.7	27.3
	7/8	22.9	23.8	29.6

From the vast number of possible options table 11.6 shows only the data for the example "QPSK embedded in 64-QAM", with $\alpha = 2$.

When comparing the data in table 11.6 with the corresponding information in table 11.5, one notices that the C/N required for QEF reception of the high-priority data stream transmitted in the QPSK constellation points, must be 3.3 dB to 4.9 dB higher than in the case of non-hierarchical modulation. Hence, the hierarchical modulation reduces the robustness of the QPSK portion. This finding should not be surprising, as the QAM constellation points lead to an actual decrease in the spacing between the apparent QPSK constellation point (the cloud of 16 points in one quadrant) and the decision thresholds. The greater the value of α the less important the effect becomes; for an increase in α results in a smaller relative size of the clouds forming the apparent QPSK constellation points.

For QEF reception of the low-priority data stream transmitted in the QAM constellation a higher C/N is again required than for the non-hierarchical 64-QAM. The differences are between 1.7 dB and 3 dB. As already explained in section 11.5, these differences are mainly due to the increase in average transmission power described by α. In the case of $\alpha = 1$ there are, in fact, only minor differences between the required C/N for non-hierarchical and hierarchical modulation (maximum 1.2 dB).

To conclude, the total effect of the introduction of the hierarchical modulation shall be discussed on the basis of a practical example ($\Delta = 1/4$). Let the hierarchical modulation be given in the form of "QPSK embedded in 64-QAM", with $\alpha = 2$. The total data stream should be receivable with a rooftop antenna (Ricean channel), while the high-priority data stream should be receivable on a stationary receiver with a rod antenna (Rayleigh channel). For the high-priority data stream, let the code rate of the inner error protection be chosen at 2/3, while for the low-priority data stream the code rate of the inner error protection be assumed to be 5/6. The data rate of 6.64 Mbit/s for the high-priority data stream can be taken from table 11.4. The required C/N is 11.7 dB. The low-priority data stream transmits 16.59 Mbit/s. For its reception a minimum C/N of 22.7 dB is required. For example, if sufficient C/N is provided in the transmission channel, a receiver with a rod antenna could receive one television programme, and a receiver with a rooftop antenna, three or four.

Symbols in Chapter 11

$g(X)$	generator polynomial defined in the time domain
I	in-phase component
i	running variable, integer
K	Ricean factor
k_{Max}	maximum value of the OFDM carrier index
k_{Min}	minimum value of the OFDM carrier index
N_e	number of echoes in a channel model
Q	quadrature component
T_g	duration of guard interval
T_U	duration of useful interval
T_S	duration of a symbol
t	time in general
X	polynomial argument defined in the time domain
$x(t)$	input signal of a channel model
$y(t)$	output signal of a channel model
z	complex-valued argument
α	distance measure of constellation points
Δ	relative length of OFDM guard interval
θ_i	phase angle in echo path i
ϱ_o	attenuation in the direct signal path
ϱ_i	attenuation in echo path i
τ_i	relative time delay in echo path i

12   Measurement Techniques for Digital Television

The measurement of characteristic parameters in digital television can be divided into two areas. First of all, the source signals in the baseband are processed to form compressed data streams. These are subsequently furnished with an error protection and then digitally modulated for transmission. The measurement techniques for the processing steps in both areas will be introduced in this chapter.

When compressing audio and video signals, noticeable changes can occur in the output signal (see chapters 3 and 4). The MPEG-2 standard offers extensive possibilities for video coding to influence the coding result. Depending on the parameters chosen and the sequential image content, the resulting image quality is subject to considerable variation. A test sequence was developed at Braunschweig Technical University with the aid of which the image quality provided by an encoder can be visually assessed. This is described in section 11.2.1.

The data streams resulting from the source coding of several programmes are combined to form a "transport stream" (TS) in the time multiplex (see chapter 5). In order to check the correctness and standard conformity of the coding of the information in the data streams, special protocol testers are required which are able to interpret and detect errors in the data.

The functionality of decoders can be checked with test data streams, which can be audio or video elementary streams or transport streams. For checking various aspects of the decoding, data streams are required which contain specific features critical for a decoder.

Digital transmission technique predominantly employs measuring methods known from other fields of communication technology. Section 12.2 will introduce several measurements obtained from transmitters and receivers, as well as measurements of the bit-error rate.

Within the DVB Project the DVB Measurement Group (DVB-MG) is currently developing the "measurement guidelines". These include tests for the MPEG-2 "transport stream" (TS) and for cable and satellite transmission. Future guidelines will also cover terrestrial transmission. With the standardisation of various measurement techniques the group hope to create a common base for content providers, network operators and manufacturers. Standardised test methods and parameters are to facilitate the implementation of a service and to help ensure smooth operation. Apart from that, the guidelines will deal with legal questions concerning contracts between broadcasters and network operators. Another goal is to ensure that measuring devices from different manufacturers can carry out the same measurements with reproducible results. A part of the work carried out to-date will be introduced in section 12.1.3. However, the standardisation procedure of the measurement guidelines has not yet been completed. Changes are therefore still possible. In the field of transmission technology the determination of certain measurements is still under discussion.

12.1 Measurement Techniques for Source-Signal Processing in the Baseband

12.1.1 Quality Evaluation of Video Source Coding

The image quality which can be obtained with hybrid coding (as, for instance, with MPEG-2) cannot at the moment be satisfactorily assessed with quantitative measurement parameters. The assessment of image quality is therefore only possible using video artefacts of a coded image sequence. This statement is also true for the comparison of encoders and decoders.

A sequence which facilitates such an assessment, and whose luminance signal can be seen in figure 12.1 [ROHDE], was developed at the Institute for Communications Technology of Braunschweig Technical University. All typical artefacts of a hybrid coding can be clearly demonstrated with this sequence. Furthermore, it enables statements to be made about the pre-processing, the structure of the "group of pictures" (GOP, see section 4.3.3) and the value of the global quantisation factor. It consists of various numbered image zones, which will be briefly described below.

Image zone 1 consists of finely shaded bar patterns and is divided into ten fields. On the left-hand side there are five fields with horizontal bars, two of which consist of black-and-white patterns, and adjacent, on the right, there are five fields with vertical bars, two of which again comprise black-and-white patterns. The remaining fields consist of green-magenta patterns, whose chrominance components C_B and C_R represent the lowest permissible values (green) and the highest permissible values (magenta) respectively. The same luminance value has been chosen for both colours in order to obtain a complete separation between luminance and chrominance for the analysis of the pre-processing. Hence the coloured bar patterns in figure 12.1 are not discernible. With the help of the above fields, one can see at a glance the steps that preceded the data compression, such as subsampling and filtering.

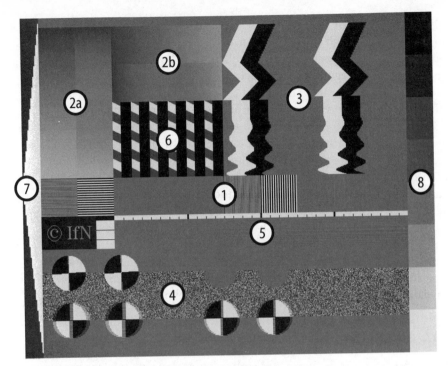

Fig. 12.1. Test sequence for the evaluation of the image quality

The vertical and horizontal grey wedges (image zones 2a and 2b) show such errors as occur due to the quantisation of the DCT coefficients of low order. Due to the brightness modulation of the grey wedges, which results in a shifting of the grey values, even negligible quantisation errors become obvious.

Image zone 3 is composed of various edges. In the left part these are immobile, while in the right part they move up and down. As before, the green and magenta coloured zones, which are adjacent to the right of the black zones, are not visible. Since the representation of the steep skirts of the edges specifically calls for DCT coefficients of a high order, but as these are only coarsely quantised at the encoder, quantisation errors are particularly prominent at the edges.

Further critical picture contents are the moving objects in image zone 4. At the edges of these objects errors are still noticeable even if – in the case of a generally good image quality – hardly any visible errors occur in the remaining picture. The sector wheels at the top on the left rotate around their centre. All others move evenly in a horizontal direction. The sector wheel comprising green and magenta at the top on the right is not identifiable for the known reasons.

Image zone 5 is for the estimation of the global quantisation factor, which can be used, by a rough approximation, as the measure of image quality and is therefore of great importance in the evaluation of the coded test sequences. It can be read off by regarding the structure border between the even grey zone on the left and the weakly structured grey zone on the right as an indicator pointing to a value between 1 and 31 on the scale above the zone.

The oblique "grey" bars (in the actual test sequence these are red) within the six vertical white stripes of image zone 6 occur from left to right with decreasing temporal frequency. In the first stripe they are to be found in every second image, whereas in the stripe furthest to the right they are only to be found in every seventh image. In conjunction with the geometric arrangement, errors caused by the prediction can be made visible in this way. In so far as these prediction errors occur obviously enough, they can be used to determine the organisation of the GOP.

Image zone 7 indicates whether, instead of a complete image, only a section of an image has been coded in order to decrease the amount of data.

Colour errors can be identified with the EBU colour bar (image zone 8).

12.1.2 Checking Compressed Audio and Video Signals

The conformance of the video and audio elementary streams (see chapters 3 and 4) with the MPEG-2 standard can be verified by means of protocol analyses. A selection of the possibilities provided by the MPEG-2 standard has been chosen by the DVB Project for audio and video coding. New, DVB-specific elements were not introduced for this level [ETR 154]. The DVB Measurement Group has therefore not dealt with this topic. Further information on testing the conformance of audio and video streams is to be found in part 4 of the MPEG-2 standard [ISO 13818], where the individual stream elements are discussed in great detail.

12.1.3 Checking the MPEG-2 Transport Stream

The MPEG-2 "transport stream" (TS) is of particular interest for protocol testing because all the useful information to be transmitted is integrated in it (see chapter 5). It is also easily accessible because the DVB Physical Interface Group (DVB-PI) has specified three different interfaces for it [EN 50083-9]:

- the data are transmitted with an 8-bit width via the synchronous parallel interface. In addition there is a clock signal, a data-valid signal and a sync signal;
- the data format of the "synchronous serial interface" (SSI) corresponds to that of the parallel interface after a parallel/serial conversion;

– the "asynchronous serial interface" (ASI) is operated with a constant data
rate of 270 Mbit/s. The jitter occurring as a result of the mapping of the useful bytes of the TS onto the high-rate stream must be equalised by the
first-in first-out (FIFO) memory at the receiver.

Errors in the source coding of the individual audio and video signals, errors
of the multiplex and some of the errors occurring during transmission can be
identified by analysing the transport stream.

A great number of tests have been recommended by the DVB Measurement Group for the MPEG-2 transport stream. These were developed for in-
service testing and can either be used for continuous or for periodic monitoring of the most important elements [ETR 290]. During the development of the
TS tests great care was taken to strike a reasonable balance between the importance of a parameter and the implementation requirements for the testing, thus making continuous monitoring of the transport stream possible
without an excessive hardware requirement.

These tests should be seen solely as a "health check" of a TS and are not intended as complete conformance tests as described in part 4 of the MPEG-2
standard [ISO 13818]. A more thorough analysis of a TS by means of further
tests is of course possible. Owing to the complexity of the TS and the higher
data rate, this is, however, not possible in real time and can therefore only be
carried out on a part of the TS. The representation in this section is limited to
the tests recommended by the Measurement Group.

Many of the tests listed below refer to the information in the TS header.
They can also be carried out with a scrambled TS (see chapters 5 and 8) since
the header always remains unscrambled. One of the prerequisites for the tests
is a quasi error-free TS as should be available under normal operating conditions at the output of the Reed-Solomon decoder. If a transmission error is
detected, or if the measuring instrument is no longer locked to the TS, further
error messages are suppressed.

The tests are conceived in such a way that particular errors are assigned to
an indicating element so that the respective protocol tester or measuring decoder can indicate the operational error to the operator in a clearly visible
form. A sequence of LEDs, labelled with the names of the errors, will suffice to
carry out a comprehensive monitoring.

The tests are divided into three groups according to their importance.
The tests in the first group check whether the TS can be synchronised and
whether no errors have occurred during transmission. In addition, errors
in the compilation of the TS can be detected. Tests in the second and third
groups are for the conformance of the repetition rates and the existence
of various elements of the TS as well as for deviations of the clock references.

The following pages refer to various syntactical elements of the MPEG-2 TS and to errors defined by the MG. In order to avoid confusion, the spelling of these elements and errors adopted here is that used in the corresponding documents [ISO 13818], [ETR 290].

Group 1: High-priority Tests (Basic Monitoring)

TS_sync_loss

The measuring device for demultiplexing and carrying out further testing must first be synchronised with the TS, for instance by means of a hysteresis function. The device can be regarded as synchronised if a certain number of correct sync bytes following upon each other at packet intervals have been received. This state of being synchronised is only changed when several consecutive sync bytes do not have the value 47_{HEX}. The MG recommends that a device be synchronised to the TS after five correct consecutive sync bytes. When there have been two or more consecutive erroneous sync bytes the synchronisation is lost.

Synchronism between the measuring device and the TS is an essential condition for all further testing. If there is no synchronism, the results of the remaining tests become irrelevant. All other error messages must therefore be suppressed or ignored.

Sync_Byte_error

If the already synchronised condition of an individual sync byte takes a value different from 47_{HEX}, then this will be indicated as Sync_Byte_error.

PAT_error

By means of the "program association table" (PAT) the demultiplexer is informed which programmes are contained in the TS and with which "packet identification" (PID) those packets are labelled in which, by means of the "program map table" (PMT), more detailed information is to be found about the compilation of the programme. PAT and PMT, therefore, have the function of an index, and it is only through these that access to the audio, video and auxiliary data is made possible.

Errors are indicated

– if packets with PID 0000_{HEX}, as specified for the PAT, do not contain any PAT data,
– if the time interval between two packets with PID 0000_{HEX} is greater than 0.5 seconds, or
– if the scrambling_control_field of the header indicates that the packet is scrambled. PAT data must not be scrambled.

Continuity_count_error

In each TS packet header there is a 4-bit field, the value of which is incremented in each packet of a particular PID. Hence, there exists one independent continuity_counter per PID. Under certain conditions a packet may be transmitted twice.

An error is identified

– if the order of the packets is not correct,
– if a packet is triply or multiply transmitted, or
– if a packet is missing.

A continuity_count_error can occur, for example, when a non-correctable transmission error is identified by the Reed-Solomon decoder. This can be indicated by the setting of the transport_error_indicator in the header of an altered packet. The packet can then no longer be unambiguously assigned to a PID because the PID itself might have been altered. It is most probable that a jump in the continuity_counter of this PID will occur in the next packet with the same PID.

Generally it must be observed that, if they are indicated by the discontinuity_indicator in the adaptation field, even discontinuities of the continuity_counter are permitted.

PMT_error

Together with the PAT described above, the "program map table" (PMT) takes care of allocating the audio, video and auxiliary data of a programme to the PIDs.

An error is indicated

– if packets with the PID of a PMT are repeated at a time interval greater than 0.5 seconds, or
– if the scrambling_control_field of the header indicates that the packet is scrambled (scrambling not being permissible in packets with PMT).

To check the conditions, the PAT must be completely decoded to start with, because it is only then that those PIDs become known with the aid of which the PMTs can be decoded.

PID_error

When the PAT and all the PMTs have been decoded, one can check, using the list of all the PIDs referenced in the tables, whether packets were actually transmitted with each PID stated in the list. A PID_error is identified when there is a PID in which no packet is transmitted within a user-defined time limit. This type of error points to an inconsistency between the PIDs of the transmitted packets, the contents of the PAT, and the PMTs. The opposite case, i.e. that of a PID not listed in the tables but appearing in the TS, is permitted and does not cause a PID_error.

Group 2: Lower-priority Tests (Recommended for Continuous or Periodic Monitoring)

Many of the following tests are only possible if the useful data of the TS are not scrambled. They sometimes require considerable computation (e.g. CRC_error), which must be balanced against the usefulness of the information gathered.

Transport_error

One bit of the TS packet header is used as the transport_error_indicator, which indicates whether uncorrected errors occurred in the packet in the course of the transmission or processing. If a bit has been set it may not be set back. The payload of the packet is unusable for decoding purposes.

If the bit is set, there is a Transport_error. Further tests can no longer deliver any reliable results. Therefore erroneous packets must not be tested.

CRC_error

A certain amount of information is transmitted in the TS in the form of tables. These include the already mentioned "program association table" (PAT) and the "program map tables" (PMTs) as well as further service information (SI) in the "network information table" (NIT), the "event information table" (EIT), the "bouquet association table" (BAT) and the "service description table" (SDT) (see section 5.4). For the transmission, the individual sections that a table is made up of are provided with a cyclic redundancy-check code. The decoder can then carry out the check and determine whether the data have been correctly received.

PCR_error

The "program clock references" (PCRs) serve to synchronise the "system time clocks" (STCs) of the multiplexer and the demultiplexer. For each programme, separate PCRs can be transmitted in adaptation fields (see section 5.3). The MPEG-2 standard stipulates that the time interval between two consecutive PCR values of the same programme must not exceed 100 ms. The DVB Implementation Guidelines [ETR 154] recommend that this interval should not be greater than 40 ms.

There is a PCR_error

- if the interval between the PCRs of a programme is greater than 40 ms, or
- if the difference between two subsequent PCRs of a programme is greater than 100 ms, without a temporal discontinuity being indicated.

To check the second condition, the coded PCRs themselves are compared, as opposed to the first condition. If the difference between them is greater than the maximum permissible interval of 100 ms, there is bound to be discontinuity in the STC. This calls for a signalling by the discontinuity_indicator in the adaptation field. If the discontinuity of the coded time is not indicated there is a PCR_error.

PCR_accuracy_error

The MPEG-2 standard [ISO 13818] specifies the tolerances for the STC. A deviation of ±810 Hz from the nominal frequency of 27 MHz is permissible. The frequency drift may be ±0.075 Hz/s at the most. In addition, an error of ±500 ns is allowed for the PCRs received. Possible causes can be inaccuracies in the generation of the PCRs. A further source of errors is the remultiplex. When new programmes are compiled for a transport stream, the PCRs must be recalculated ("restamping") due to the fact that their position in the data stream can be changed. If the data of a programme are repeatedly subjected to a remultiplex, care should be taken that the PCRs do not exceed the maximum permissible deviations.

Discrepancies in the reception time of the PCRs due to jitter on the transmission path must not be registered for these tolerances. Thus, carrying out these tests only requires knowledge of the PCRs and of their position in the TS. The reception time is of no importance here.

As each programme can have its own time base, the PCRs of several programmes may have to be checked separately. Alternatively, the test can also be carried out for one chosen programme.

An inequality for checking the conformance with the specifications is to be found in part 4 of the MPEG-2 standard. However, this method can only be applied if the data rate in the TS is constant.

PTS_error

The "presentation time stamps" (PTSs) supply the synchronisation of the video and audio signals of a programme. The stipulated maximum time interval, which should not exceed 700 ms in any elementary stream, is to be checked. Due to the fact that the PTS values are in the header of the "packetised elementary stream" (PES) they are only available with an unscrambled data stream.

CAT_error

The information required for decoding scrambled transport streams is coded in the "conditional-access table" (CAT).

A CAT_error is indicated

- if a TS contains packets whose transport_scrambling_control fields signal scrambled packets although there is no CAT, or
- if a section of another table is coded in a packet with PID 0001_{HEX}, reserved for the CAT.

Group 3: Optional Tests (Application-dependent Monitoring)

In this group the majority of the tests refer to further tables with service information [ETS 468]. These include the Network Information Table_error (NIT_error), the Service Description Table_error (SDT_error), the Event Information Table_error (EIT_error), the Running Status Table_error (RST_error), and the Time and Date Table_error (TDT_error), each of which is checked as to whether the packets of the PID defined by the DVB Project contain other tables than those specified. The point to be taken into consideration in this context is that the DVB Project has defined a stuffing table which may be contained in service information packets beside the useful information. In addition to the above procedure, adherence to the specified minimum repetition rates in the tables is tested. These rates vary, depending on the tables [ETR 211].

For the Service Information_repetition_error (SI_repetition_error), in addition to checks being carried out on the minimum repetition rates already included in the above individual tests, the sections of the SI are tested as to whether they are not repeated too fast.

For the Buffer_error the course of the buffer occupancy of the ideal decoder described in the MPEG-2 standard ("transport stream system target decoder" [T-STD]) has to be reconstructed. An overflow of one buffer, also an underflow of several buffers, will cause a Buffer_error.

A TS may also contain streams with PIDs which are not listed in the PAT or in a PMT. If this is the case, it is indicated as Unreferenced_PID.

12.1.4 Checking the Functionality of the Decoder

A further field of measurement techniques is the checking of the conformity of decoders with the standards. Here, there can be problems for two reasons. One reason is the great flexibility of the MPEG-2 standard. For example, it is acceptable, but not recommended, to code a data stream with plenty of overhead in the form of headers and time stamps which a decoder might not be able to decode with sufficient speed. The second reason lies in the fact that the MPEG-2 standard [ISO 13818] describes an ideal decoder, the T-STD, which, in the form described, cannot be implemented because, among other things, in the T-STD the data are removed from the interim stores instantaneously and not over a period of time. Consequently, manufacturers of decoders are bound to choose an implementation which varies from the ideal decoder, but one which nevertheless enables a decoding conforming with the standard.

For checking the correct functioning of decoders it is appropriate to use special test streams, by means of which different coding aspects can be realised for different purposes. For example, a change in the resolution and in the aspect ratio can be coded in a video elementary stream, or all the possible motion vectors can be coded in a data stream (cf. section 4.2.2). The same applies to the audio coding and the multiplex.

In the development stage of the MPEG-2 standard, video and audio elementary streams were coded and exchanged by the participating companies and organisations in order to verify the standard.

Transport streams for testing purposes can contain special video and audio streams, or they show peculiarities of the multiplex. Examples in this field are:

- a change of the programme definition by the coding of new tables (PAT and PMT),
- a jump of the STC (cf. section 5.3),
- the coding of the PCR with the maximum permissible deviations.

For production and service, data streams with which the tasks of contemporary resolution-test-pattern generators can be accomplished are also of interest. For even in the future the test patterns familiar from analogue television will be required for controlling cathode ray tubes. They can be coded to form video streams and can then be multiplexed to form a transport stream. The test patterns fed with such a TS could be shown on a DVB receiver and one could switch between images with programme selection buttons.

The supply of a DVB receiver with the necessary data streams will require a generator. This is composed of a storage medium for the data streams, an interface for the output of the data streams, and a device for the control of the operational run and for the internal data transport.

12.2 Measurements for Digital Transmission Technology

An overview of some important measurement quantities and their respective measuring instruments for the transmission technology of digital television is given in the following table:

Measurement quantities	Measuring instruments
eye diagram, ISI	oscilloscope
vector diagram (constellation)	vector signal generator or vector modulation analyser
power distribution	power analyser
linearity	power analyser, vector modulation analyser
interference; S/N, C/N ratio	spectral analyser, special measuring instruments
error rate	transmission analyser, additional decoder for on-line measurement
regenerator phase noise	phase detector, spectrum analyser

Interference, for instance expressed as S/N or C/N power ratio, considerably affects the achievable error rate of the transmission. As is sufficiently known, the said powers can be determined by a spectrum analyser. For in-service measurements it is possible to use a channel-state analyser, which can carry out an analysis on the basis of the behaviour of the forward error correction (FEC, see chapter 6) when the source signals are known.

The following sections describe the measurement of the error rate and the power distribution as well as the eye diagram and vector diagram, followed by the measurement of the regenerator phase noise.

12.2.1 Representation of the Eye Diagram

The eye diagram presents a very clear measure of the quality of a digital transmission [SÖDER]. Generally, this measure is the size of the eye-opening area, although the amplitude and phase noise as well as the clock jitter complicate the practical evaluation. The eye diagram usually only presents an overall evaluation of the system which allows no conclusions about individual errors.

The eye diagram is very easy to represent by means of an oscilloscope as shown by the block diagram of the measurement set-up in figure 12.2. In a binary system the decision threshold and its sampling instant correspond with the symmetry axes of the eye diagram.

Intersymbol interference (ISI, see section 7.1) occurs in the transmitted signal when the previous or subsequent signals affect the actual signal in its eye opening. Examples are shown in figure 12.2. When the impulse response of the entire transmission channel shows zeros for integral multiples of the clock period the system is free from intersymbol interference (first Nyquist criterion), not, however, from overshoots which could cause overdrive effects and consequently lead to a non-linear behaviour of the channel. This point was discussed in detail in section 7.1.

For multilevel systems the eye opening decreases in proportion to the number of decisions. Figure 12.2 shows a quaternary system with four ampli-

Representation of an eye diagram   **Examples**

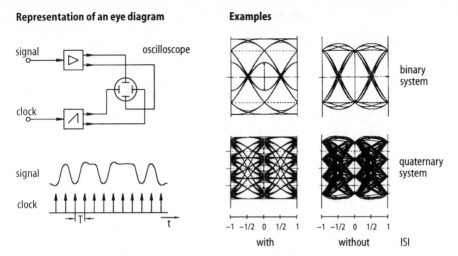

Fig. 12.2. Eye diagrams: representation and examples [SÖDER]

tude thresholds and three eye openings. The decision levels are at the centre line of the respective eye level (dash-dot lines).

12.2.2 Measurements Carried out at Modulators and Demodulators

The principle of quadrature amplitude modulation (QAM) has already been explained in detail in section 7.5. Figure 12.3 shows as an example the block diagram of a QAM transmission. The mapper (byte-to-symbol-word converter) divides the incoming serial data stream into two parallel signals I and Q (in-phase and quadrature phase), which, subsequent to passing through low-pass filters, modulate in quadrature (90° phase shift to each other) a carrier signal in its amplitude. The D/A converters required for multilevel I- and Q-signals are assumed to be included in the mapper.

Subsequent to running through the transmitter amplifier, the transmission path and the receiver input stages, the QAM signal is demodulated in the reverse order. For this purpose, a reference carrier signal has to be regenerated from the received signal and is then fed to both the synchronous demodulators. Another requirement is that the phase position of this reference signal be known in relation to the modulation. A full account of this procedure has been given in section 10.4. The stages of the signal processing, as described here, are assumed to be integrated in the carrier regenerator of the block diagram.

The block diagrams of the transmitter and the receiver, as illustrated here, are also valid for a vector signal generator or for a vector modulation analyser [SCHERER]. Both measuring instruments contain the same structural com-

Transmitter   **Path**

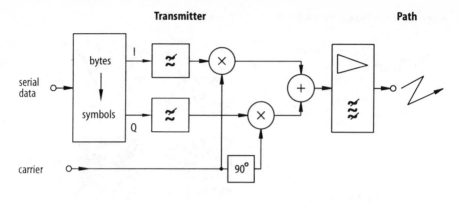

Path   **Receiver**

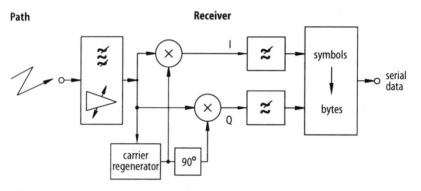

Fig. 12.3. Principle of the quadrature modulation

ponents as represented here, that is, with reference to their functions. However, the components are designed as universally usable sets with very close
tolerances. With these measuring sets it is possible to carry out detailed investigations not only of whole systems, but also of individual components. This
is the subject of the following considerations, which generally pertain to other
types of modulation as well, e.g. the modulation with quadrature phase shift
keying (QPSK) which can be represented as QAM with binary I- and Q-
signals.

12.2.2.1 Measurements at a QAM transmitter

Figure 12.4 shows a measurement set-up. The eye diagrams of the baseband
signals can be represented by means of a vector modulation analyser, for instance in order to investigate the effect of the band-limiting low-pass filters.
In the carrier-frequency domain the vector representation of the signal components is useful in the (complex) I,Q-plane since it facilitates the measuring
of static or dynamic amplitude- and phase deviations. In this case, three rep-

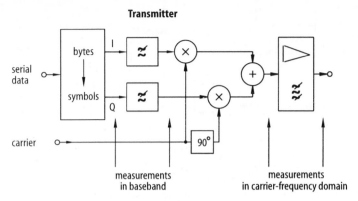

Fig. 12.4. Measurements carried out at a QAM transmitter

resentations are possible: firstly, one which continuously shows the whole temporal course with all transients but which often has an unclear effect; secondly, a temporally discrete representation of the vector diagram which only indicates the optimal signal-sampling instants ("constellation", see section 7.5); and, thirdly, the temporal representation of the signal according to its amplitude and phase course.

The power analyser measures the average and the maximum power of the modulated signal [SCHERER, RHODES]. Theoretically there is a relationship between the two values which is given by the type of modulation (peak-to-average power ratio); however, in practice this relationship can be changed by non-linearities in the transmitter. The display of a histogram of the output power for the various vectors of the QAM makes a further investigation into this effect possible.

12.2.2.2 Measurements at a QAM Receiver

Figure 12.5 shows the construction of a suitable measurement set-up. For the purpose of measuring the characteristics of the QAM receiver by means of a universal, practically ideal transmitter it is recommendable to use a vector signal generator for the input. This generator enables definite changes of various system parameters, e.g. of the I/Q amplitude ratio, the quadrature angle, or the characteristics of the I/Q pre-filters. Depending on these parameter variations, the respective deviations in the receiver can be measured with a vector modulation analyser as described above. Sometimes the answer required is the error rate, the measurement of which is explained in section 12.2.3.

The simulation of the transmission path comprises its partition into individual parallel channels, which can be equipped independently of each other with varying attenuation, delay time, and Doppler shift [SCHERER]. By com-

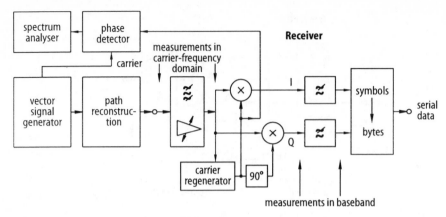

Fig. 12.5. Measurements carried out at a QAM receiver

bining these individual channels a transmission path of considerable complexity is created which, controlled by programme, is also dynamically variable. In this way both multipath propagation and fading effects can be simulated; the fading effects may, in addition, show a correlation between the individual simulation channels.

For measuring the properties of the carrier regenerator, especially its phase noise, a phase detector is provided which reads out the phase difference between the very stable carrier signal of the vector signal generator and the regenerated reference signal of the QAM receiver. The phase noise and other interference in the reference signal can thus be detected with a spectrum analyser. The above-mentioned properties have a direct effect on the width of the opening in the eye diagram at the receiving end.

12.2.2.3 Measurements at an OFDM Transmission

The transmission of digital signals by means of an orthogonal frequency division multiplex (OFDM), which is also referred to as multiple-carrier technique, has already been described in section 7.6. Firstly, the measurements at an OFDM transmission system in the carrier-frequency domain include the testing of the individual subcarriers for their correct frequency and modulation. This is possible by means of a spectrum analyser with suitable resolution if a blockwise periodically recurring data pattern is transmitted (cf. section 11.3.3). Secondly, it is also possible to measure, with the same equipment, the complete power density in the channel. In order to determine the power and the overdriving limits of the transmitter, the respective distribution of the output signal is also important. This can be ascertained by means of a power analyser, as described in section 12.2.2.1. What was stated in section 12.2.2.2 is also valid for the simulation of the transmission path.

12.2.3 Measurement of Error Rate

The error rate in a digital transmission is the most frequently used quality criterion. However, this is only an overall measure of quality, which barely expresses anything about the individual causes of errors. Contrary to the error probability which was often used as operand in the preceding sections, the error rate is a measured value whose determination in its turn will lead to further errors. Therefore there can only be an approximate conformity between the two quantities.

12.2.3.1 Direct Measurement of Error Rate

The measuring arrangement shown in figure 12.6 comprises a generator which creates a pseudo-random binary sequence (PRBS) that is fed into the

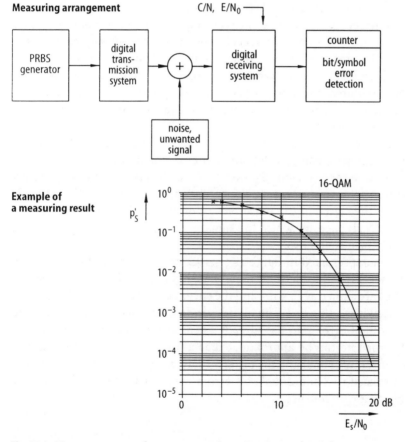

Fig. 12.6. Direct measurement of error rate: measuring arrangement and example

digital transmission system as a data stream. The interference which is super-imposed on the transmission path is represented as the addition of noise and interference signals in the block diagram. In the digital reception system the decisions as to which signs or symbols were transmitted are thus corrupted, the corruption being the result of a random process. These decisions are checked at the bit/symbol-error identification stage and the errors found are counted, either for a predetermined duration or for a specified number of decisions [WELLHSN].

The duration of such a measurement is therefore dependent on these base values, on the error rate itself, and on the permissible measuring error. It can be very long for a statistically relevant number of errors if their occurrence is infrequent. Furthermore, this measuring technique requires either a free channel or implementation during non-operational times, so that it is mainly used in laboratories or when putting a system into operation.

The length of the pseudo-random sequence should be chosen to be as long as possible, so that the shape of its spectrum is almost uniform for all data rates used. To recognise errors at the receiving end, in the case of link tests, it is necessary to synchronise the receiver with the known sequence of the transmitter by cross-correlation. However, the duration of this procedure is directly dependent on the length of the sequence, so that a compromise between both requirements is necessary. For the known hierarchies of the digital transmission systems with high data rates, sequences of the length of $2^{15} - 1$ and $2^{23} - 1$ are given.

The error rate is usually represented as dependent on the signal-to-noise ratio. In the carrier-frequency domain of the system this quantity is the ratio of the carrier power to the noise power C/N, which is relatively easy to determine with a spectrum analyser. In the baseband, the quantity is S/N, which is the ratio of the signal to the noise power; or else the ratio of the energies E_b/N_o or E_s/N_o is used, for which the signal energy refers to a bit or a symbol respectively (see chapter 7).

The curve represented in figure 12.6 serves as an example of a measuring result in a 16-QAM transmission (16 possible modulation vectors). The measured values are shown as crosses, and the curve calculated in accordance with (7.46) is shown as a thin line. The interdependence between the bit-error rate and the symbol-error rate has already been described in section 7.4. The symbol-error rate can be approximated as a product of the symbol length (here, for instance, 4 bits) and the bit-error rate, as long as the last-mentioned does not exceed 10^{-3} and double bit errors within a symbol are practically negligible as a consequence.

Measuring arrangement

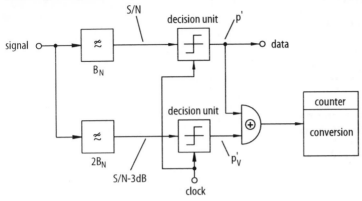

Fig. 12.7. Comparison measurement of error rate

12.2.3.2 Comparison Measurement of Error Rate

Figure 12.7 shows the measuring arrangement for a method that permits the comparison measurement of the error rate even during operation [FEHER 1, FEHER 2]. This method is valid on the assumption that the impairing interference and its effect on the transmission system are known, as indicated by a curve like the one shown in figure 12.6.

For this method there are two decoders of the same type present in the receiver and both are fed with the same signal. The only difference between the two branches is the noise bandwidth of the pre-switched channel filters, the latter are symbolised here by low-pass filters. The noise bandwidth of the comparison channel is to be chosen double that of the data channel. It follows from this that the S/N ratio in the comparison channel – assuming a spectrally uniformly distributed interference – will be 3 dB less than in the data channel.

On the further assumption that the error rate in the data channel is much smaller than that in the comparison channel, the data channel can be used as an "ideal" reference for the evaluation of the errors in the comparison channel. This evaluation takes place by means of an EX-OR gate, at the output of which the discrepancies between the two decision units are counted. By conversion, using the predetermined error-rate characteristic of the system, the corresponding value for the data channel is obtained.

Due to the considerably greater error rate in the comparison channel the measuring speed increases accordingly. Variations of this method which have become known include the feeding of an additional interference signal into the comparison channel and the changing of the threshold value belonging to that channel.

Symbols in Chapter 12

B_N	noise bandwidth
E_b	energy/bit
E_s	energy/symbol
I	in-phase component
N_o	noise-power density
p'	error rate
p'_S	symbol-error rate
p'_V	comparative error rate
Q	quadrature component
T	clock period
t	time in general

13 Bibliography

[AGRAW]   Agrawal, B.P.; Shenoi, K.: Design Methodology for M. IEEE Trans. on Comm., COM-31, No 3, March 1983, pp 360–370

[APPELQ]   Appelquist, P.: HD-DIVINE, a Scandinavian terrestrial HDTV project. EBU Technical Review, No 256, Summer 1993, pp 16–19

[BENVEN]   Benveniste, A.; Goursat, M.: Blind Equalizers. IEEE Trans. on Comm. COM-32, No 8, August 1984, pp 871–883

[BEYERS]   Beyers, B.W.; Christofer, L.; Saint Girons, R.; Teichner, D.: DirecTV/ USSB/DSS – das erste Multi-Kanalsystem für die digitale Satellitenüber-tragung von Fernsehprogrammen. 16. Jahrestagung der Fernseh- u. Ki-notechn. Gesellschaft (FKTG), Nürnberg 1994, proceedings, p 227

[BIAESCH]   Biaesch-Wiebke, C.: CD-Player und R-DAT-Recorder. 3rd edition. Vogel Buchverlag, Würzburg 1992

[BIDET]   Bidet, E.; Joanblanc, C.; Senn, P.: A Fast Single Chip Implementation of 8192 Points Complex FFT. Proc. of the CMOS Integrated Circuits Conference (CICC), San Diego 1994

[BRONST]   Bronstein, I.N.; Semendjajew, K.A.: Taschenbuch der Mathematik. Verlag Harri Deutsch, Thun 1984

[CANDY]   Candy, J.C.: A Use of Double Integration in Sigma Delta Modulation. IEEE Trans. on Comm., COM-33, No 3, March 1985, pp 249–258

[CCIR 10]   Multi-channel stereophonic sound system with and without accompanying picture. CCIR Draft New Recommendation (Doc. 10/11), Doc. 10/BL/3. March 1992

[CCIR 563] Radiometeorological Data. CCIR Report 563-4, 1990

[CCIR 564] Propagation Data and Prediction Methods Required for Earth-Space Telecommunication Systems. CCIR Report 564-4, 1990

[CLARK]   Clark, G.C.; Cain, J.B.: Error-Correction Coding for Digital Communications. Plenum Press, New York 1988

[COSTAS]   Costas, J.P.: Synchronous Communications. Proc. IRE 44 (1956), pp 1713–1718

[DE BOT]   de Bot, P.G.M.; Le Floch, B.; Mignone, V.; Schütte, H.D.: An Overview of the Modulation and Channel Coding Schemes Developed for Digital Terrestrial Television Broadcasting within the dTTb Project. Proc. of the Internat. Broadcasting Convention, Amsterdam 1994, p 569

[DIGISMATV] Satellite Digital TV in Collective Antenna Systems. SMATV Reference Channel Model for Digital TV. Race Digital Satellite Master Antenna Television (DIGISMATV) Project, Doc. DIGISMATV-NT-PART (A-D)-031-HSA, June 1994

[EN 50083-9]   Cabled distribution systems for television, sound and interactive multi-media signals; Part 9: Interfaces for CATV/SMATV headends and similar professioal equipment for DVB/MPEG-2 transport streams. European Norm EN 50083-9, Comité Européen de Normalisation Électrotechnique CENELEC, June 1998

[EN 50221]   Common Interface Specification for Conditional Access and other Digital Video Broadcasting Decoder Applications. European Norm EN 50221, Comité Européen de Normalisation Électrotechnique CENELEC, February 1997

[EN 748]   Digital Video Broadcasting (DVB); Multipoint Video distribution Systems (MVDS) at 10 GHz and above. European Norm EN 300 748, European Telecommunications Standards Institute ETSI, August 1997

[EN 749]   Digital Video Broadcasting (DVB); Framing structure, channel coding and modulation for Multipoint Video Distribution Systems (MVDS) below 10 GHz. European Norm EN 300 749, European Telecommunications Standards Institute ETSI, August 1997

[ENGELS]   Engels, V.; Rohling, H.; Breide, S.: OFDM-Übertragungsverfahren für den Digitalen Fernsehrundfunk. Rundfunktech. Mitt. 37 (1993), No 6, pp 260–270

[ETR 154]   Digital Video Broadcasting (DVB); Implementation guidelines for the use of MPEG-2 systems, Video and Audio in satellite, cable and terrestrial broadcasting applications. ETSI Technical Report ETR 154, European Telecommunications Standards Institute ETSI, October 1997

[ETR 211]   Digital Video Broadcasting (DVB); Allocation of Service Information (SI) codes for DVB systems. ETSI Technical Report ETR 211, European Telecommunication Standard ETSI, August 1997

[ETR 289]   Digital Video Broadcasting (DVB); Support for use of scrambling and Conditional Access (CA) within digital broadcasting systems. ETSI Technical Report ETR 289, European Telecommunications Standards Institute ETSI, October 1996

[ETR 290]   Digital Video Broadcasting (DVB); Measurement guidelines for DVB systems. ETSI Technical Report ETR 290, European Telecommunications Standards Institute ETSI, August 1997

[ETS 401]   Radio broadcasting systems: Digital Audio Broadcasting (DAB) to mobile, portable and fixed receivers. European Telecommunication Standard ETS 300 401, European Telecommunications Standards Institute (ETSI), 1994

[ETS 421]   Digital Video Broadcasting (DVB); DVB framing structure, channel coding and modulation for 11/12 GHz satellite services. European Norm 300 421, European Telecommunications Standards Institute ETSI, October 1997

[ETS 429]   Digital Video Broadcasting (DVB); DVB framing structure, channel coding and modulation for cable systems. European Norm 300 429, European Telecommunications Standards Institute ETSI, April 1994

[ETS 468]   Digital Video Broadcasting (DVB); Specification for Service Information (SI) in Digital Video Broadcasting (DVB) Systems. European Norm 300 468, European Telecommunications Standards Institute ETSI, February 1998

[ETS 472]   Digital Video Broadcasting (DVB); Specification for conveying ITU-R System B Teletext in Digital Video Broadcasting (DVB) bitstreams. European Norm 300 472, European Telecommunications Standards Institute ETSI, August 1997

[ETS 473]   Digital Video Broadcasting (DVB); Satellite Master Antenna Television (SMATV) distribution systems. European Norm 300 473, European Telecommunications Standards Institute ETSI, August 1997

[ETS 743]   Digital Video Broadcasting (DVB); Subtitling systems. European Technical standard ETS 300 743, European Telecommunications Standards Institute ETSI, September 1997

[ETS 744]   Digital Video Broadcasting (DVB); Framing structure, channel coding and modulation for digital terrestrial television (DVB-T). Draft European Norm EN 300 744, European Telecommunications Standards Institute ETSI, January 1999

[FCC]   Federal Communications Commission: ATV System Recommendation. IEEE Trans. on Broadcasting 39 (1993), No 1, pp 2–208

[FEHER 1]   Feher, K.: Digital Communications: Microwave Applications. Prentice-Hall Inc., Englewood Cliffs, N.J. 1981

[FEHER 2]   Feher, K. u.a.: Telecommunications Measurements, Analysis and Instrumentation. Prentice-Hall Inc., Englewood Cliffs, N.J. 1987

[FLAHERTY]   Flaherty, J.A.: Digital Television and HDTV in America – A progress report. EBU Technical Review, No 260, Summer 1994, pp 30–35

[FORNEY]   Forney, G.D.: Burst-Correcting Codes for the Classic Bursty Channel. IEEE Trans. on Comm. Technol. COM-10, No 5, October 1971, pp 772-781

[FRIEDR]   Friedrich, J.; Herbig, P.; Reuber, H.J.: Theorie und Praxis bandbreiteneffizienter Modulations- und Codierungverfahren. ANT Nachrichtentechn. Berichte, No 11, March 1994, pp 23–40

[FTZ 1]   Bezugskette für die Übertragung von Fernseh- und Tonrundfunksignalen im nationalen Breitbandkommunikations(BK)-Verteilnetz. FTZ-Richtlinie 151R8, Fernmeldetechnisches Zentralamt, Darmstadt 1985

[FTZ 2]   Bedingungen und Empfehlungen für den Anschluß privater Breitband/Rundfunk-Empfangsantennenanlagen. FTZ-Richtlinie 1R8-15, Fernmeldetechnisches Zentralamt, Darmstadt 1985

[FURRER]   Furrer, F.J.: Fehlerkorrigierende Block-Codierung für die Datenübertragung. Birkhäuser Verlag, Basel 1981

[GAUSS]   Gauß, C.F.: Theorie der den kleinsten Fehlern unterworfenen Combinationen der Beobachtungen. In: Boersch, A.; Simon, P. (Hrsg): Abhandlungen zur Methode der kleinsten Quadrate. Stankiewicz Buchdruckerei, Berlin 1887, pp 1–27

[GERDSEN]   Gerdsen, P.; Kröger, P.: Digitale Signalverarbeitung in der Nachrichtenübertragung. Springer-Verlag, Berlin Heidelberg New York 1993

[GRALLI]   Grand Alliance HDTV System: Specification Version 2.0. Chapter 6: Transmission System. Chapter 7: Grand Alliance System Summary. 7 December 1994

[GUILLOU]   Guillou, L.C.; Giachetti, J.L.: Encipherment and Conditional Access. SMPTE Journal 103 (1994), No 6, pp 398–406

[HAGENAUER]   Hagenauer, J.: Rate-Compatible Punctured Convolutional Codes (RCPC Codes) and their Applications. IEEE Trans. on Comm. COM-36, No 4, April 1988, pp 389-400

[HESSENM]   Hessenmüller, H.; Jakubowski, H.: Die Signalqualität bei digitaler
               Tonübertragung – subjektive Testergebnisse, objektive Meßverfahren.
               Rundfunktech. Mitt. 27 (1983), No 1, pp 3–16
[HOPKINS]   Hopkins, R.: Digital terrestrial TV for North America – The Grand Alli-
               ance HDTV System. EBU Technical Review, No 260, Summer 1994, pp
               36–50
[HUNG]   Hung, A.C.: PVRG-MPEG CODEC 1.1. Portable Video Research Group
               (PVRG), Stanford University, June 1993
[IDSMITTEN]   In der Smitten, F.J.: Digital Video Broadcasting – Feldversuche zur digi-
               talen Übertragung OFDM-codierter Farbbildsignale in einem terrestri-
               schen Fernsehkanal. Rundfunktech. Mitt. 37 (1993), No 3, pp 130–141
[IEC 958]   Interface audionumérique/Digital audio interface. Norme Internatio-
               nale/International Standard CEI/IEC 958,1989-03. Commission Électro-
               technique Internationale/International Electrotechnical Commission,
               Geneva 1989
[IRT]   Richtlinie für die Beurteilung der Fernsehversorgung bei ARD/ZDF und
               DBP. Institut für Rundfunktechnik, Richtlinie No 5R10, Munich 1981
[IRWIN]   Irwin, D.W.; Cox, N.R.: Hybrid Clamping in NTSC Digital Video Equip-
               ment. IEEE Trans. on Broadcasting 38 (1992), No 1, pp 19–26
[ISO 10918]   Information technology – Digital compression and coding of con-
               tinuous-tone still images. ISO/IEC International Standard IS 10918, July
               1992
[ISO 11172]   Information technology – Coding of moving pictures and associated
               audio for digital storage media up to about 1.5 Mbit/s. ISO/IEC Interna-
               tional Standard IS 11172, November 1992
[ISO 13818]   Information technology – Generic coding of moving pictures and asso-
               ciated audio. ISO/IEC International Standard IS 13818, November 1994
[ISO 7498]   Information processing systems – Open Systems Interconnection – Ba-
               sic Reference Model. ISO International Standard IS 7498, 1984
[ITU 601]   Studio Encoding Parameters of Digital Television for Standard 4:3 and
               Wide-Screen 16:9 Aspect Ratios. Draft Revision of Recommendation
               ITU-R BT.-601-4. International Telecommunication Union, Geneva 1994
[ITU 656]   Interfaces for Digital Component Video Signals in 525-line and 625-line
               Television Systems Operating at the 4:2:2 level of Recommendation 601.
               ITU-R Recommendation BT.656-1 (originally CCIR Rec. 656-1, Geneva
               1986)
[ITU H.261]   Video codec for audiovisual services at p x 64 kbit/s. ITU-T Recommen-
               dation H.261, March 1993
[ITU H.263]   Video coding for low bit rate communication. ITU-T Recommendation
               H.263, February 1998
[JAKUBOW]   Jakubowski, H.: Quantisierungsverzerrungen in digital arbeitenden
               Tonsignalübertragungs- und -verarbeitungssystemen. Rundfunktech.
               Mitt. 24 (1980), No 2, pp 91–92
[JANSKY]   Jansky, D.M.: Methods for Accommodation of HDTV Terrestrial Broad-
               casting. IEEE Trans. on Broadcasting 37 (1991), No 4, pp 152–157
[JOHANN]   Johann, J.: Modulationsverfahren. Nachrichtentechnik, vol 22. Sprin-
               ger-Verlag, Berlin Heidelberg New York 1992
[KAMMEYER]   Kammeyer, K.D.: Nachrichtenübertragung. Verlag B.G. Teubner, Stutt-
               gart 1992

[KENTER]   Kenter, H. (Hrsg): Ton- und Fernsehübertragungstechnik und Technik
                leitergebundener BK-Anlagen. R.v.Decker's Verlag, Heidelberg 1988
[KNOLL]   Knoll, A.: Der MPEG-2-Standard zur digitalen Codierung von Fernseh-
                signalen. Der Fernmeldeingenieur, Verlag f. Wissenschaft u. Leben G.
                Heidecker, Erlangen, July 1992
[LAFLIN]   Laflin, N.J.; Wright, D.T.: Planning and Field Testing for Digital Terres-
                trial Television. Proc. of the Internat. Broadcasting Convention, Amster-
                dam 1994, p 527
[LEE]   Lee, E.A.; Messerschmitt, D.G.: Digital Communication. 2nd edition. Klu-
                wer Academic Publishers, Boston-Dordrecht-London 1994
[LOHSCH]   Lohscheller, H.: A Subjectively Adapted Image Communication System.
                IEEE Trans. on Comm. COM-32, No 12, December 1984, pp 1316–1322
[LÜKE 1]   Lüke, H.D.: Signalübertragung – Grundlagen der digitalen und analo-
                gen Nachrichtenübertragungssysteme. 5th edition. Springer-Verlag, Ber-
                lin Heidelberg New York 1992
[LÜKE 2]   Lüke, H.D.: Korrelationssignale. Springer-Verlag, Berlin Heidelberg
                New York 1992
[MÄUSL]   Mäusl, R.: Digitale Modulationsverfahren. 3. Auflage. Telekommunika-
                tion, Bd 2. Hüthig Buch-Verlag, Heidelberg 1991
[MORGENST]   Morgenstern, G.: Zur Berechnung der spektralen Leistungsdichte von
                digitalen Basisband-Signalen. Der Fernmelde-Ingenieur 33 (1979), No
                12, pp 1–39
[MUSCHALL]   Muschallik, C.: Einfluß der Oszillatoren im Frontend auf ein OFDM-
                Signal. Fernseh- und Kino-Technik 49 (1995), No 4, pp 196–205
[MUSMANN]   Musmann, H.G.; Pirsch, P.; Grallert, H.D.: Advances in Picture Coding.
                Proc. IEEE 73 (1985), No 4, pp 523–548
[NHK]   Nippon Hoso Kyokai (NHK): Digital Modulation Scheme for ISDB (Inte-
                grated Services Digital Broadcasting) in the 12 GHz Band. NHK Labora-
                tories Note No 428, September 1994
[NORTH]   North, D.O.: An Analysis of the Factors which Determine Signal/Noise
                Discrimination in Pulsed-Carrier Systems. Proc. IRE 51 (1963), No 7. pp
                1016–1027
[NYQUIST]   Nyquist, H.: Certain Topics in Telegraph Transmission Theory. Trans. of
                the American Institute of Electrical Engineers 47, February 1928, pp
                617–644
[OPPENHM]   Oppenheim, A.V.; Willsky, A.S.: Signale und Systeme. VCH-Verlag,
                Weinheim 1989
[PARK]   Park, S.: Principles of Sigma-Delta Modulation for Analog-to-Digital
                Converters. Application Note APR8/D, Motorola Inc. 1990
[PENNEBK]   Pennebaker, W.B.; Mitchell, J.L.: JPEG still image data compression
                standard. Van Nostrand Reinhold, New York 1993
[PETKE]   Petke, G.: Planungsaspekte für digitales terrestrisches Fernsehen. Rund-
                funktech. Mitt. 37 (1993), No 3, pp 119–129
[PROAKIS]   Proakis, J.G.: Digital Communications. Second Edition. McGraw-Hill
                Book Company, New York 1989
[REED]   Reed, S.; Solomon, G.: Polynomial Codes over Certain Finite Fields. Jour-
                nal of the Society for Industrial and Applied Mathematics 8, June 1960,
                pp 300-304

[REIMERS 1]   Reimers, U.: European perspectives on digital television broadcasting –
               Conclusions of the Working Group on Digital Television Broadcasting
               (WGDTB). EBU Technical Review, No 256, Summer 1993, pp 3–8
[REIMERS 2]   Reimers, U.: Systemkonzepte für das Digitale Fernsehen in Europa.
               Fernseh- und Kino-Technik 47 (1993), No 7–8, pp 451–461
[REIMERS 3]   Reimers, U.: Das europäische Systemkonzept für die Übertragung digi-
               talisierter Fernsehsignale per Satellit. Fernseh- und Kino-Technik 48
               (1994), No 3, pp 115–123
[REIMERS 4]   Reimers, U.: Concept of a European System for the Transmission of Dig-
               itized Television Signals via Satellite. SMPTE Journal 103 (1994), No 11, pp
               741–747
[REIMERS 5]   Reimers, U.: Digitales Fernsehen für Europa – Ein Statusbericht. Fern-
               seh- und Kino-Technik 48 (1994), No 10, pp 517–519
[RHODES]   Rhodes, C.W.: Measuring Peak and Average Power of Digitally Modu-
               lated Advanced Television Systems. IEEE Trans. on Broadcasting 38, No
               4, December 1992, pp 197–201
[RODDY]   Roddy, D.: Satellitenkommunikation. Hanser Verlag, München-Wien /
               Prentice Hall, London 1991
[ROHDE]   Encoder Test-Sequenz. Produktinformation DVTS PD 757.1790.21,
               Rohde & Schwarz GmbH & Co. KG, Munich, June 1995
[ROY]   Roy, A.; Reimers, U.: FDMA-Betrieb eines Satelliten-Transponders für
               die Zuführung mehrerer TV-Programme zu terrestrischen PAL-
               Sendern. Rundfunktechn. Mitt. 39 (1995), No 2, pp 63–70
[RUELBG]   Ruelberg, K.D.: Kanalcodierung und Modulation für digitale Fern-
               sehübertragung über Satellit, Kabel und terrestrische Sendernetze. 16.
               Jahrestagung der Fernseh- und Kinotechn. Gesellschaft (FKTG), Nürn-
               berg 1994, proceedings, pp 139–162
[SCHAAF]   Schaaf, C.: Digital Delivery by Cable. IEE Symposium on Emerging
               Broadcast Technology, Cambridge 1994
[SCHERER]   Scherer, K.: Measurement Tools for Digital Video Transmission. IEEE
               Trans. on Broadcasting 39, No 4, December 1993, pp 350–363
[SCHÖNFD 1]   Schönfelder, H. (Hrsg): Digitale Filter in der Videotechnik. Drei-R-
               Verlag, Berlin 1988
[SCHÖNFD 2]   Schönfelder, H.: Fernsehtechnik. Teil 2. Justus von Liebig-Verlag, Darm-
               stadt 1973
[SCHÖPS]   Schöps, G.: Bandbreiteneffiziente Modulationsverfahren für die Ka-
               belübertragung von Digital-HDTV. Rundfunktechn. Mitt. 38 (1994), No
               4, pp 60–67
[SCHÜSSLER]   Schüßler, H.W.: Digitale Signalverarbeitung. Band I: Analyse diskreter
               Signale und Systeme. Springer-Verlag, Berlin Heidelberg New York 1988
[SKRITEK]   Skritek, P.: Handbuch der Audio-Schaltungstechnik. Franzis-Verlag,
               Munich 1988
[SÖDER]   Söder, G.; Tröndle, K.: Digitale Übertragungssysteme. Theorie, Optimie-
               rung und Dimensionierung der Basisbandsysteme. Springer-Verlag,
               Berlin Heidelberg New York Tokyo 1985
[STENGER]   Stenger, L.: Das Systemkonzept für die Übertragung digitaler Fernseh-
               signale im Kabel. 16. Jahrestagung der Fernseh- und Kinotechn. Ge-
               sellschaft (FKTG), Nürnberg 1994, proceedings, pp 67–80

[STOLL 1]   Stoll, G.; Theile, G.: MASCAM: Minimale Datenrate durch Berücksichti-
               gung der Gehöreigenschaften bei der Codierung hochwertiger Tonsig-
               nale. Fernseh- und Kino-Technik 42 (1988), pp 551 -558

[STOLL 2]   Stoll, G.: MPEG-2 Layer 2 "Musicam-Surround": The new multichannel
               audio coding standard for broadcast, telecommunication and multime-
               dia. 16. Jahrestagung der Fernseh- und Kinotechn. Gesellschaft (FKTG),
               Nürnberg 1994, proceedings, pp 101–113

[SWEENEY]   Sweeney, P.: Codierung zur Fehlererkennung und Fehlerkorrektur. Han-
               ser Verlag, München-Wien / Prentice Hall, London 1992

[TEICHNER]   Teichner, D.; Herpel, C.; Schröder, E.F.; Spille, J.; Riemann, U.: Der
               MPEG-2-Standard – Generische Codierung für Bewegtbilder und zu-
               gehörige Audio-Information. Fernseh- und Kino-Technik 48 (1994), Nos
               4–10

[TELEKOM]   Preliminary Criteria for Planning Digital Terrestrial Television Services.
               Deutsche Bundespost Telekom, ITU-R Document 11C/28-E, 28 July 1993

[THEILE]   Theile, G.; Stoll, G.; Link, M.: Low bitrate coding of high quality audio sig-
               nals – An introduction to the MASCAM system. EBU Technical Review,
               No 230, August 1988, pp 158–181

[TIETZE]   Tietze, U.; Schenk, Ch.: Halbleiter-Schaltungstechnik. 10. Auflage. Sprin-
               ger-Verlag, Berlin Heidelberg New York 1993

[TS 101 197-1] Digital Video Broadcasting (DVB); DVB Simulcrypt; Part 1: Head-end
               architecture and synchronization. Technical Specification TS 101 197-1.
               European Telecommunications Standards Institute ETSI, June 1997

[UNGER]   Unger, H.G.: Elektromagnetische Wellen auf Leitungen. 2nd edition.
               Hüthig-Verlag, Heidelberg 1986

[VELDERS]   Velders, A.M.; Tichelaar, J.Y.: Measurement on PALplus Signals in Dutch
               CATV Networks. Dutch National Platform HDTV, Hilversum 1993

[VITERBI]   Viterbi, A.J.: Error Bounds for Convolutional Codes and an Asymptoti-
               cally Optimum Decoding Algorithm. IEEE Trans. on Information The-
               ory IT-13 (1967), No 2, pp 260-269

[WALLACE]   Wallace, G.K.: The JPEG Still Picture Compression Standard. IEEE Trans.
               on Consumer Electronics CE-38, No 1, February 1992

[WELLHSN]   Wellhausen, H.W.; Martin, D.: Fehlerhäufigkeitsmessungen. Nachrich-
               tentechn. Zeitschrift 24, Heft 11, November 1971, pp 553–558

[WGDTB]   Reimers, U. et al.: Report to the European Launching Group on the Pros-
               pects for Digital Terrestrial Television. Europäisches DVB-Projekt, Do-
               cument WGDTB 1063, Geneva, November 1992

[ZANDER]   Zander, H.: Die digitale Audiotechnik. Grundlagen und Verfahren. 1st
               edition. Drei-R-Verlag, Berlin 1987

[ZWICKER]   Zwicker, E.: Psychoakustik. Springer-Verlag, Berlin 1982

14 Acronyms and Abbreviations

ABSOC   Advanced Broadcasting Systems of Canada
AC   alternating-current (component)
A/D   analogue-to-digital
ADC   analogue-to-digital converter
ADSL   Asymmetrical Digital Subscriber Line
AES   Audio Engineering Society
AF   adaptation field
AFC   automatic frequency control
AGC   automatic gain control
AM   amplitude modulation
ARD   Arbeitsgemeinschaft der öffentlich-rechtlichen Rundfunkanstalten der
         Bundesrepublik Deutschland (one of the German public broadcasters)
ASK   amplitude shift keying
ATM   asynchronous transfer mode
ATV   advanced television
AWGN   additive white Gaussian noise

BAT   bouquet association table
BCH   Bose-Chaudhuri-Hocquenghem (code)
BER   bit-error rate
BP   band-pass filter
BSS   broadcasting-satellite service
BTA   Broadcasting Technology Association (Japanese industrial organisa-
         tion)

CA   conditional access
CAT   conditional-access table
CATV   community-antenna television
CCIR   Comité Consultatif International de Radiodiffusion (now ITU-R)
CCITT   Comité Consultatif International du Télégraphe et du Téléphone
         (now ITU-T)
CD, CD-I compact disc, compact disc – interactive
CEC   Commission of the European Communities
CEI/IEC Commission Électrotechnique Internationale/International Electro-
         technical Commission
CENELEC Comité Européen de Normalisation Électrotechnique
CEPT   Conférence Européenne des Postes et des Télécommunications
C/I   carrier-to-interference ratio (in dB)
CLK   clock (timing signal)

CMOS	complementary metal-oxide semiconductor
C/N	carrier-to-noise ratio (in dB)
CNR	carrier-to-noise ratio = C/N
COFDM	coded orthogonal frequency-division multiplex
CRC	cyclic redundancy check
CVBS	colour-video blanking synchronisation
D2-MAC	television transmission standard (sound and data: digital [duobinary]; image: multiplexed analogue component)
DA	distribution amplifier in broadband cable plants
D/A	digital-to-analogue
DAB	Digital Audio Broadcasting
DAC	digital-to-analogue converter
DAPSK	differential amplitude phase-shift keying
DAVIC	Digital Audio Visual Council
DBP	Deutsche Bundespost (former Federal German postal authority)
DC	direct-current (component)
DCO	digitally controlled oscillator
DCT	discrete cosine transform
DFT	discrete Fourier transform
DPCM	differential pulse-code modulation
DSC	digital serial components
DSR	digital satellite radio
DSS	digital satellite system
DTS	decoding time stamp
DTVB	Digital Television Broadcasting
DVB	Digital Video Broadcasting
DVB-MC	DVB Microwave Cable-based
DVB-MG	DVB Measurement Group
DVB-MS	DVB Microwave Satellite-based
DVB-PI	DVB Physical Interface
DVC	digital video cassette
DVD	Digital Versatile Disc
EAV	end of active video
EBU	European Broadcasting Union = EUR
ECM	entitlement control message
EDTV	enhanced-definition television
EIRP	equivalent isotropically radiated power
EIT	event information table
EMM	entitlement management message
EN	European Norm
ERO	European Radiocommunications Office
ES	elementary stream
ETR	ETSI Technical Report
ETS	European Telecommunication Standard
ETSI	European Telecommunications Standards Institute
EX-OR	exclusive-or (function) = XOR
FCC	Federal Communications Commission
FDMA	frequency-division multiple access

FEC	forward error correction
FFT	fast Fourier transform
FIFO	first in – first out (memory)
FKTG	Fernseh- und Kinotechnische Gesellschaft (German Society for Television, Film and Electronic Media)
FM	frequency modulation
FSK	frequency-shift keying
FSS	fixed-satellite service
FTZ	Fernmeldetechnisches Zentralamt (former German central office of telecommunications)
GF	Galois field
GOP	group of pictures
HD-MAC	high-definition MAC
HDTV	high-definition television
HE	head end of broadband cable plants
HEX	hexadecimal
HP	high-pass filter
IBO	input back-off
IC	integrated circuit
ID	identification, identifier
IDCT	inverse discrete cosine transform
IDFT	inverse discrete Fourier transform
IEC	International Electrotechnical Commission
IEEE	Institute of Electrical and Electronics Engineers
IF	intermediate frequency
I/Q	in-phase/quadrature phase
IRT	Institut für Rundfunktechnik (research centre of public broadcasting organisations [in Austria, Germany and Switzerland])
ISDN	integrated services digital network
ISI	intersymbol interference
ISO	International Standardization Organization
ITU	International Telecommunication Union = UIT
ITU-R	International Telecommunication Union – Radiocommunication
ITU-T	International Telecommunication Union – Television
JPEG	Joint Photographic Experts Group
LDTV	low-definition television
LED	light-emitting diode
LF	loop filter (of PLL)
LNB	low-noise block
LP	low-pass (filter or signal)
LSB	least significant bit
MAC	multiplexed analogue component
MDCT	modified discrete cosine transform
MHP	multimedia home platform
MMDS	microwave multichannel/multipoint distribution system
MP@ML	main profile at main level

MPEG	Moving Pictures Experts Group
MR	multiresolution
MR-QAM	multiresolution QAM
MSB	most significant bit
MUSE	Multiple Subsampling Encoding (Japanese analogue high-definition television system)
MUX	multiplex, multiplexer
NAB	National Association of Broadcasters
NHK	Nippon Hoso Kyokai (Japanese broadcasting corporation)
NIT	network information table
NRZ	non-return to zero
NTSC	National Television Systems Committee
OBO	output back-off
OFDM	orthogonal frequency division multiplex
OMUX	output multiplexer
OSI	open systems interconnection
PACF	periodically recurring autocorrelation function
PAL	phase alternating line
PALplus	enhanced PAL system
PAT	program association table
PCM	pulse-code modulation
PCR	program clock reference
PD	phase discriminator
PDH	plesiochronous digital hierarchy
PDM	pulse-density modulation
PES	packetised elementary stream
PH	PES header
PID	packet identification
PLL	phase-locked loop
PMT	program map table
PRBS	pseudo-random binary sequence
PS	program stream
PSD	power spectral density
PSI	program-specific information
PSK	phase-shift keying
PTS	presentation time stamps
PVRG	Portable Video Research Group
QAM	quadrature amplitude modulation
QEF	quasi error-free
QPSK	quadrature phase-shift keying
RF	radio frequency
ROM	read-only memory
RS	Reed-Solomon (code)
RST	running-status table
SAR	successive approximation register
SAV	start of active video

SAW	surface acoustic wave
SCR	system clock reference
SDT	service description table
SDTV	standard-definition television
SECAM	Séquentiel couleur à mémoire (French-developed analogue colour TV system)
SER	symbol-error rate
S&H	sample-and-hold (circuit)
SI	service information
SIF	source input format
(S)MATV	(satellite) master antenna television
SMPTE	Society of Motion Picture and Television Engineers
SMS	subscriber management system
S/N	signal-to-noise ratio (in dB)
SNR	signal-to-noise ratio
STC	system time clock
STERNE	Système de télévision en radiodiffusion numérique
TA	trunk amplifier in broadband cable plants
TA_A	trunk amplifier of network level A in broadband cable plants
TA_B	trunk amplifier of network level B in broadband cable plants
TDT	time and data table
TH	transport-stream header
TM	Technical Module (of the DVB Project)
TP	transition point
TPS	transmission parameter signalling
TS	transport-stream
T-STD	transport-stream system target decoder
TWTA	travelling wave tube amplifier
UER	Union Européenne de Radiodiffusion = EBU
UHF	ultra-high frequencies (470–862 MHz, television)
UIT	Union Internationale des Télécommunications = ITU
VBS	video blanking synchronisation
VCO	voltage-controlled oscillator
VHF	very high frequency (47–300 MHz, television)
VHS	video home system (standard for video recorders)
VLC	variable-length coder
VLD	variable-length decoder
VSB	vestigial-sideband amplitude modulation
WGDTB	Working Group on Digital Television Broadcasting
XOR	exclusive-or (function) = EX-OR
ZDF	Zweites Deutsches Fernsehen (one of the German public broadcasters)
ΔM	delta modulator
$\Sigma\Delta M$	sigma-delta modulator

15   Index

Printing: Mercedesdruck, Berlin
Binding: Buchbinderei Lüderitz & Bauer, Berlin